edexcel

REVISE EDEXCEL GCSE

Science

REVISION WORKBOOK

Higher

Series Consultant: Harry Smith
Series Editor: Penny Johnson

Authors: Peter Ellis, Damian Riddle, Julia Salter, Ian Roberts

THE REVISE EDEXCEL SERIES

Available in print or online

Online editions for all titles in the Revise Edexcel series are available Autumn 2012.

Presented on our ActiveLearn platform, you can view the full book and customise it by adding notes, comments and weblinks.

Print editions

Science Revision Workbook Higher 9781446902622

Science Revision Guide Higher 9781446902615

Online editions

Science Revision Workbook Higher 9781446904763

Science Revision Guide Higher 9781446904749

Print and online editions are also available for Science (Foundation), Additional Science (Higher and Foundation) and Extension Units.

This Revision Workbook is designed to complement your classroom and home learning, and to help prepare you for the exam. It does not include all the content and skills needed for the complete course. It is designed to work in combination with Edexcel's main GCSE Science 2011 Series.

To find out more visit:
www.pearsonschools.co.uk/edexcelgcsesciencerevision

Contents

A small bit of small print

Target grade ranges are quoted in this book for some of the questions. Students targeting this grade range should be aiming to get most of the marks available. Students targeting a higher grade should be aiming to get all of the marks available.

Edexcel publishes Sample Assessment Material and the Specification on its website. This is the official content and this book should be used in conjunction with it. The questions in this book have been written to help you practise what you have learned in your revision. Remember: the real exam questions may not look like this.

Classification

E-C

1 Bacteria are unicellular organisms. Each bacterium is made up of one cell only.

bacteria amoeba

Amoebae are also unicellular organisms.

Bacteria are placed into the kingdom Prokaryotae, whereas green algae are placed into the kingdom Protoctista. Suggest one reason why bacteria and amoebae are placed in different kingdoms.

....

....

(2 marks)

E-C

2 Describe two reasons why animals and plants are placed in separate kingdoms.

....

....

(2 marks)

C-B

3 Explain why scientists do not classify viruses in any of the five kingdoms.

Think about how viruses differ from other organisms and use the word 'because' in your answer.

....

....

....

(3 marks)

C-B Guided

4 **a)** The table shows how some organisms are classified. Complete the table to give the correct classification group or binomial name for the organisms shown.

Classification group	Humans	Wolf	Panther
	Animalia	Animalia	Animalia
	Chordata	Chordata	Chordata
Class	Mammalia	Mammalia	Mammalia
Order	Primate	Carnivora	Carnivora
Family	Hominidae	Canidae	Felidae
Genus	Homo	Canis	Panthera
	Sapiens	Lupus	Pardus
Binomial name	Homo sapiens	*Canis lupus*	*Panthera pardus*

This is a guided question. This means some extra help has been given. In this case the binomial classification has been done for you.

(4 marks)

b) Explain which two organisms in the table are most closely related.

....

....

(2 marks)

Vertebrates and invertebrates

D-C 1 Use the information in the passage to answer the questions that follow.

Dolphins are homeotherms that live in oceans. They rise to the surface of the water about once every 28 seconds to breathe air into their lungs. Dolphin eggs are fertilised inside the body and the female dolphin gives birth to live young. Dolphins are covered in fine hair and produce milk to feed their young.

Dolphin

Red-eyed tree frog

The red-eyed tree frog has moist, smooth skin that can absorb oxygen. It is also able to breathe through its lungs on land. The red-eyed tree frog depends on the temperature of its surroundings to control its body temperature. It lays eggs, which are then fertilised by the male outside the female's body.

Guided

a) Suggest why the dolphin and the red-eyed tree frog are placed into different groups even though they are both vertebrates.

Dolphins are different from red-eyed tree frogs because they ....

.... and .... **(2 marks)**

b) Give one piece of evidence from the passage that suggests that the red-eyed tree frog is a poikilotherm.

....

....

(1 mark)

c) Explain which of the organisms is oviparous.

....

....

(2 marks)

D-C 2 State how organisms in the phylum Chordata are similar.

....

(1 mark)

EXAM ALERT

Vertebrates are animals with backbones. Make sure that you say 'backbone' rather than 'spinal cord', which is the nervous tissue inside the backbone.

Students have struggled with exam questions similar to this – **be prepared!** ResultsPlus

B-A 3 Explain why it is difficult to classify some vertebrates based on their reproductive methods.

....

....

(2 marks)

Species

D-C Guided

1 Place a tick in the table if you think the description is correct for a species. Place a cross in the table if you think the description does not apply to a species.

Unable to produce offspring	Show variation	Interbreed to produce fertile offspring	Have only a few features in common
✗			

(4 marks)

D-C

2 Use the key to identify the organisms found in leaf litter. Write the scientific names of the organisms underneath their picture. Three organisms have already been identified for you.

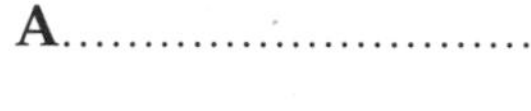

A.............................

B Rhopalomesites tardyi

C.............................

D Pterostichus melanarius

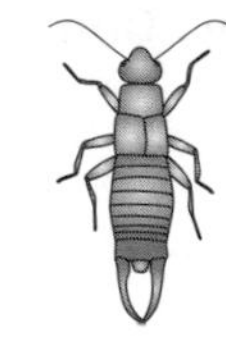

E.............................

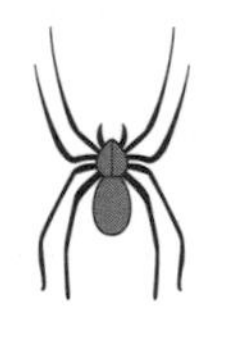

F Linyphia triangularis

(3 marks)

Don't worry about the long names – the question is testing your ability to use the key.

1	4 pairs of legs fewer than 4 pairs of legs or no legs	**go to 2** **go to 3**
2	body in two parts body in one part	*Linyphia triangularis* *Leiobunum vittatum*
3	body divided into many segments body divided into few segments	*Lumbricus terrestris* **go to 4**
4	pincers on a tail no pincers on a tail	*Euborellia annulipes* **go to 5**
5	body covered with spots no spots present	*Coccinellidae Chilocorinae* **go to 6**
6	pointed nose blunt head	*Rhopalomesites tardyi* *Pterostichus melanarius*

3 A scientist has discovered a beetle that she thinks is a new species. Suggest some of the stages she will have to go through before the scientific community agrees that she has discovered a new species of beetle.

....

....

....

(3 marks)

Binomial classification

C-A **1** Honey badgers (*Mellivora capensis*), North American badgers (*Taxidea taxus*), Eurasian badgers (*Meles meles*), stink badgers (*Mydaus javanensis*) and ferret badgers (*Melogale personata*) are all members of the same mammalian family, but are otherwise not closely related to each other.

a) Name the genus to which the ferret badger belongs.

....

(1 mark)

b) Using evidence from the passage suggest why the common names for the badgers are misleading.

> This is a two mark question, so you need to make two points in your answer.

....

....

(2 marks)

C-A **2** Explain why a binomial classification system for each living organism is useful to scientists.

....

....

(2 marks)

B-A* Guided **3** Using the example of ducks, explain why is it difficult to classify some organisms as a particular species.

Similar species of duck can breed .... to produce ....

which shows ....

....

(3 marks)

B-A* **4** There are many species of salamander. Some scientists think that these salamanders all come from one common species and that they have formed a ring species.

Describe how salamanders can provide an example of a ring species.

....

....

....

(3 marks)

> You will have learned about ring species in your lessons. You may not have used salamanders as an example but the same principles apply to all ring species.

Reasons for variety

D-C

EXAM ALERT

1 The graphs give information about height and blood group in a population of people.

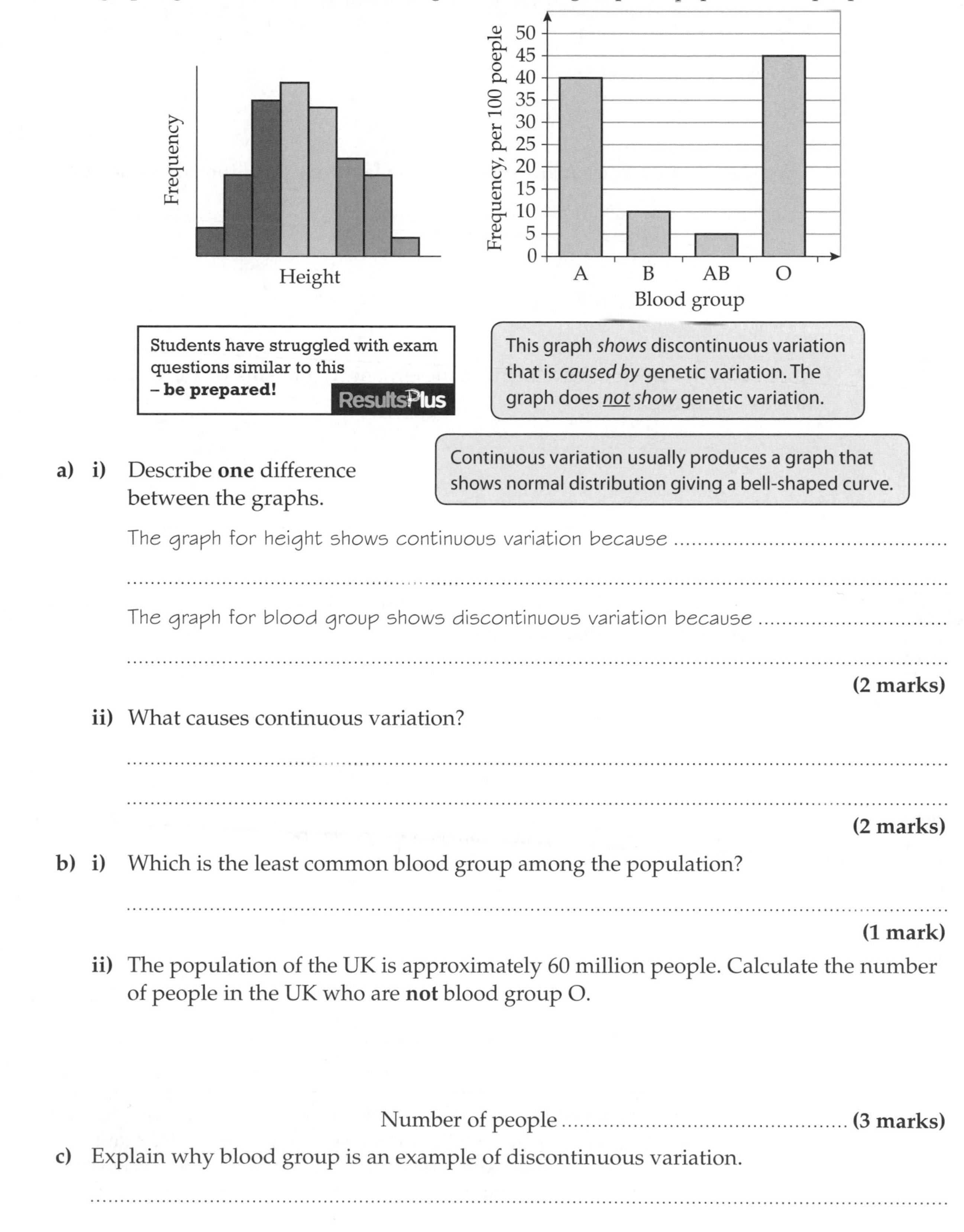

Guided

a) i) Describe **one** difference between the graphs.

The graph for height shows continuous variation because ....

....

The graph for blood group shows discontinuous variation because ....

....

(2 marks)

ii) What causes continuous variation?

....

....

(2 marks)

D-C

b) i) Which is the least common blood group among the population?

....

(1 mark)

B-A*

ii) The population of the UK is approximately 60 million people. Calculate the number of people in the UK who are **not** blood group O.

Number of people .... **(3 marks)**

c) Explain why blood group is an example of discontinuous variation.

....

....

(1 mark)

d) State the cause of discontinuous variation in a characteristic such as blood group.

....

(1 mark)

Evolution

D-C

1 a) State **two** resources that organisms compete for in an environment.

....

....

....

(2 marks)

b) Some species produce many offspring. Explain why this is an advantage to the species.

....

....

....

(2 marks)

C-A

2 Explain why, when an environment changes, some organisms within a species survive whereas others die.

....

....

....

(2 marks)

C-A

Guided

3 When a new species is discovered, a scientist may take some of its DNA to analyse. Suggest two reasons why a scientist might analyse the DNA of a new species.

To help .... the new species and to find out which other

....

....

(2 marks)

C-A*

4 Describe how geographic isolation can lead to speciation.

....

....

....

....

....

(4 marks)

A common mistake is to discuss organisms *adapting* to a new environment. This is not the case with geographic isolation. You should discuss organisms that already have advantageous characteristics when explaining speciation by geographical isolation.

Genes

E-C

1 Describe the structure of the nucleus.

....

....

(2 marks)

C-A

2 People with brown or blue eyes have different combinations of two alleles. One recessive allele codes for blue eyes and one dominant allele codes for brown eyes.

> The genotype of an individual is always represented by two letters. A dominant allele is given a capital letter (for example, B), and the recessive allele is given the lower case letter (for example, b).

Guided

a) Explain what is meant by the term 'alleles'. Use the letters above for eye colour to help you with this.

Alleles are different .... of ....

One is .... (represented by B), and one is .... (represented by b).

(4 marks)

EXAM ALERT

b) i) State the heterozygous genotype for eye colour.

....

(1 mark)

> Students have struggled with exam questions similar to this **– be prepared!** ResultsPlus

> Be careful not to confuse the meanings of *heterozygous* and *homozygous*.

ii) Explain why the person with a heterozygous genotype for eye colour will have brown eyes.

....

....

....

(2 marks)

iii) A girl has blue eyes. Explain what the girl's genotype must be.

....

....

(2 marks)

> You need to know what dominant, recessive, genotype, phenotype, homozygous and heterozygous mean.

Explaining inheritance

C-B

1 Two plants both have the genotype Tt. The two plants are bred together.

The allele that makes the plants grow tall is represented by T, and the allele that makes plants shorter is represented by t.

a) Complete the Punnett square to give the gametes of the parents and the genotypes of the offspring.

EXAM ALERT

Exam questions similar to this have proved especially tricky in the past – **be prepared!** ResultsPlus

Take great care to complete the square correctly and use the right letters.

(2 marks)

Guided

b) What percentage of the offspring from this cross will be short? Explain how you know this.

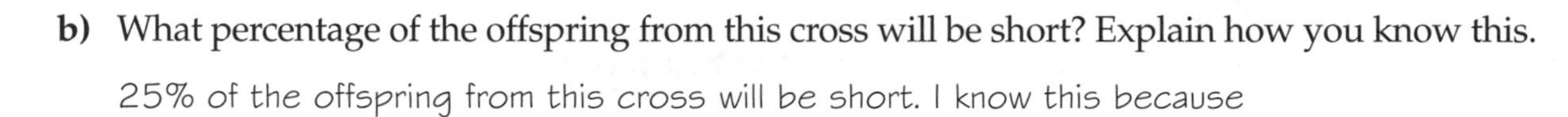

25% of the offspring from this cross will be short. I know this because

....

....

(3 marks)

c) Calculate the probability of the offspring from this cross being tall.

....

....

(1 mark)

Percentage probabilities from Punnett squares will always be 0, 25%, 50%, 75% or 100%, depending on the number of squares with a particular genotype (0, 1, 2, 3 or 4 squares). In fractions, probabilities will always be 0, $\frac{1}{4}$, $\frac{1}{2}$, $\frac{3}{4}$ or 1.

B-A*

Guided

2 Fur colour in mice is represented by two alleles, G and g. Two parent mice were bred, and produced four offspring. 50% of the offspring were white, which is the recessive characteristic. Complete the genetic diagram to show this cross and show the genotypes of the parents.

Parent genotypes	....		....	
Gametes	....	....	....	....
Genotype of offspring	....	....	....	....
Phenotype of offspring	Grey	White	Grey	White

(4 marks)

With this question it might be easier to start with what you know – the phenotypes of the offspring – and then work backwards.

Genetic disorders

C-A **1** The family pedigree shows the inheritance of cystic fibrosis.

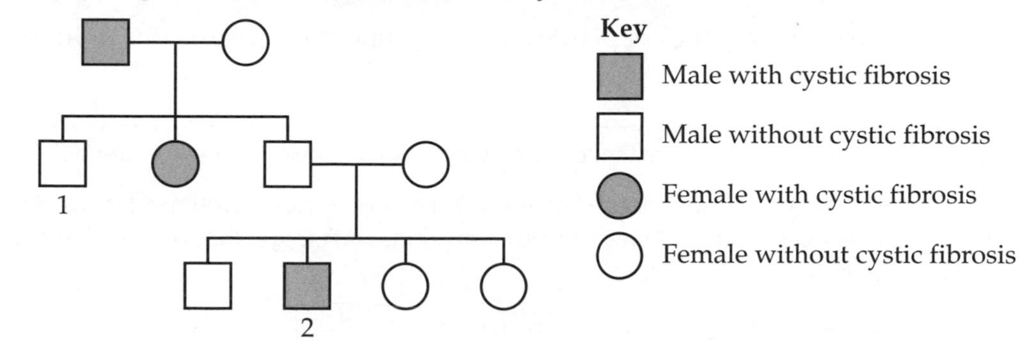

Cystic fibrosis is caused by a recessive allele. An individual must inherit two faulty alleles to show the symptoms of cystic fibrosis.

a) How many females in the family pedigree suffer from cystic fibrosis?

....

(1 mark)

b) How many males in the family pedigree have a homozygous recessive genotype?

....

(1 mark)

Guided

c) State the genotype of person 1 and explain your answer.

Person 1 does not have cystic fibrosis. This means that they must have one .... allele from their mother. But they must have inherited a .... allele from their father. This means that their genotype is ....

(3 marks)

Person 1 can only have one particular genotype. You can only work these out from looking at the genotypes of the parents. Decide what alleles each parent will pass on before you attempt to answer this question.

d) Explain how the genotype of person 2 is different from the genotype of his brother and two sisters.

....

....

....

(2 marks)

C-A **2** Describe how pedigree analysis can be useful for families with a history of sickle cell disease.

....

....

....

(2 marks)

Biology extended writing 1

Plants and fungi both play a role in recycling carbon in the environment.

Compare the characteristics of plants and fungi, and the roles they have in the carbon cycle.

(6 marks)

You will be more successful in extended writing questions if you plan your answer before you start writing.

The question asks you to compare characteristics of plants and fungi. You need to point out similarities *and* differences between the two things you are comparing. Don't forget that you need to think about the roles of these organisms in the carbon cycle.

The features that you should think about when answering this question are:

- the components of plant and fungal cells
- the way in which they make their food
- the way they use or make carbon dioxide.

Remember to use appropriate scientific language in your answer – to describe the processes where carbon dioxide is used up or made, and the ways in which the organisms make their food.

If parts of your answer contain something about only one of the kingdoms then think about what you can write about the other kingdom to match that point.

Biology extended writing 2

The peacock is a bird that usually lives in India. Male peacocks have a very colourful tail, made up of many feathers. Male peacocks show their tails to attract females.

Suggest how and why the peacock has evolved this tail.

(6 marks)

You will be more successful in extended writing questions if you plan your answer before you start writing.

The question asks you to explain the evolution of a peacock's tail. Think about:

- Why does a peacock need a tail in the first place – what are peacocks competing for?
- What happens when a peacock has a large tail and what impact does this have on the genes in the population?
- What would happen to smaller peacocks who could not put on such a display?
- What happens over time in the population?

Although time is limited in any examination, it's worth spending just a little time to think about a question and to look for the key information. Here, the key information to look for includes:

- only **male** peacocks have this tail
- the picture shows how large the tail is; and the question says it is colourful
- the question tells us that the tail has developed through the process of **evolution**.

Homeostasis

D-C **1** State the meaning of the term 'homeostasis'.

....

....

(1 mark)

D-C **2** More sweat is lost on a hot day compared with a cold day. Describe how this will affect the volume and concentration of urine.

....

....

(2 marks)

C-A **3** An investigation was carried out into the effect of exercise on core (internal) body temperature and the temperature of the skin's surface. The graph shows the results of this investigation.

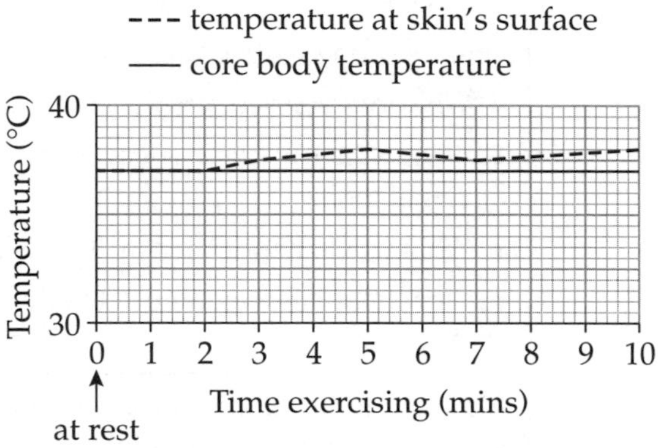

--- temperature at skin's surface
— core body temperature

Temperature (°C): 30, 40
Time exercising (mins): 0, 1, 2, 3, 4, 5, 6, 7, 8, 9, 10
at rest

a) Compare the trends in the data shown by the graph.

....

....

(2 marks)

EXAM ALERT

Guided

b) Explain how blood vessels in the skin cause a rise in the temperature of the skin's surface during exercise.

When the temperature rises blood vessels

....

There is greater ....

....

This means that more ....

(3 marks)

Students have struggled with exam questions similar to this – **be prepared!** ResultsPlus

It is important to use scientific words such as 'radiation' and 'vasodilation'.

c) Explain the why it is important that the core body temperature remains stable during exercise.

....

....

(2 marks)

C-A **4** Describe the role of the nervous system in thermoregulation.

....

....

(2 marks)

Sensitivity

E-C **1** Name **two** organs of the body that make up the central nervous system.

....

....

(2 marks)

C-A **2** The diagram shows a motor neurone.

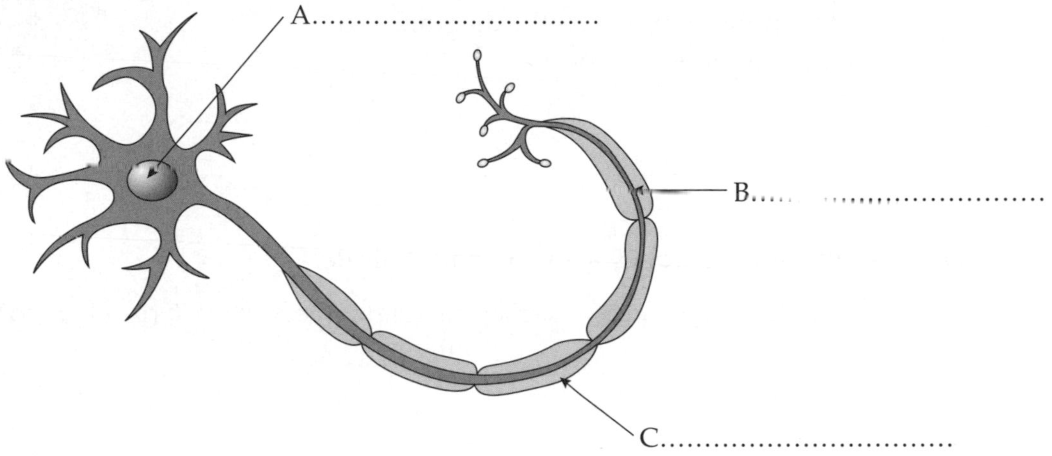

A...............................

B..... ....

C...............................

a) i) Name parts A, B and C of the motor neurone. Write your answers on the diagram.

(3 marks)

ii) Describe the roles of structures A, B and C.

A ....

B ....

C ....

(3 marks)

Don't say that neurones carry 'messages'. You have to be more specific and talk about electrical impulses.

Guided **b)** Describe **one** way that the structure of a sensory neurone differs from a motor neurone.

The cell body of a sensory neurone is ....

The cell body of a motor neurone is ....

(2 marks)

c) Describe how the function of a sensory neurone differs from the function of a motor neurone.

....

....

....

....

(4 marks)

Responding to stimuli

C-A **1** The diagram shows a junction where neurone X meets neurone Y.

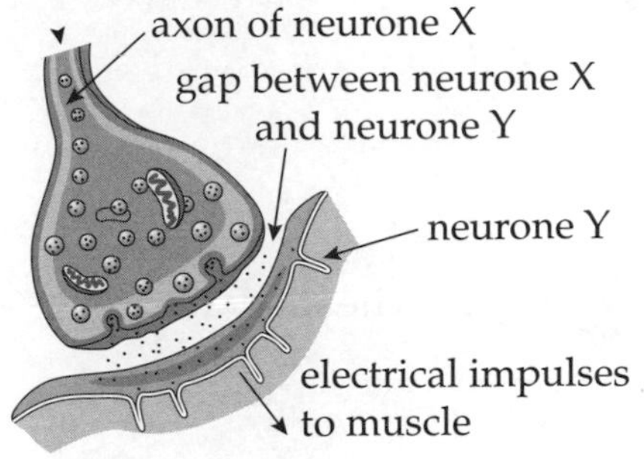

electrical impulse
axon of neurone X
gap between neurone X and neurone Y
neurone Y
electrical impulses to muscle

a) State the name given to the junction between two neurones.

....

(1 mark)

b) Label the motor neurone on the diagram and give a reason for your choice.

....

....

(2 marks)

Guided **c)** Describe how neurones X and Y communicate.

When an electrical impulse reaches the end of neurone X it travels across the

.... by ....

(3 marks)

C-A **2** The diagram shows a reflex arc.

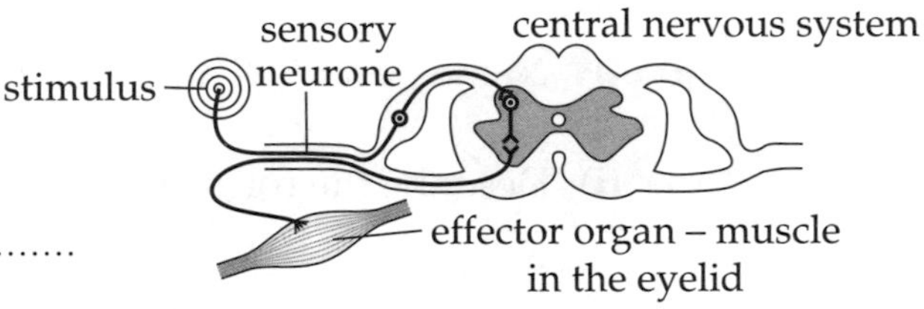

stimulus
sensory neurone
central nervous system
effector organ – muscle in the eyelid

Explain what the stimulus is in the reflex arc shown in the diagram.

....

....

....

....

(2 marks)

Make sure that you follow all the command words. For example, here the question asks you to 'explain'. This often means you need the word 'because' in your answer.

B-A* **3** **a)** When your hand touches a hot object a reflex pulls your hand away quickly. Describe the pathway taken by a nerve impulse during a reflex action.

....

....

....

....

(4 marks)

b) Describe the function of the reflex arc and how its structure helps it to carry out this function.

....

....

....

(3 marks)

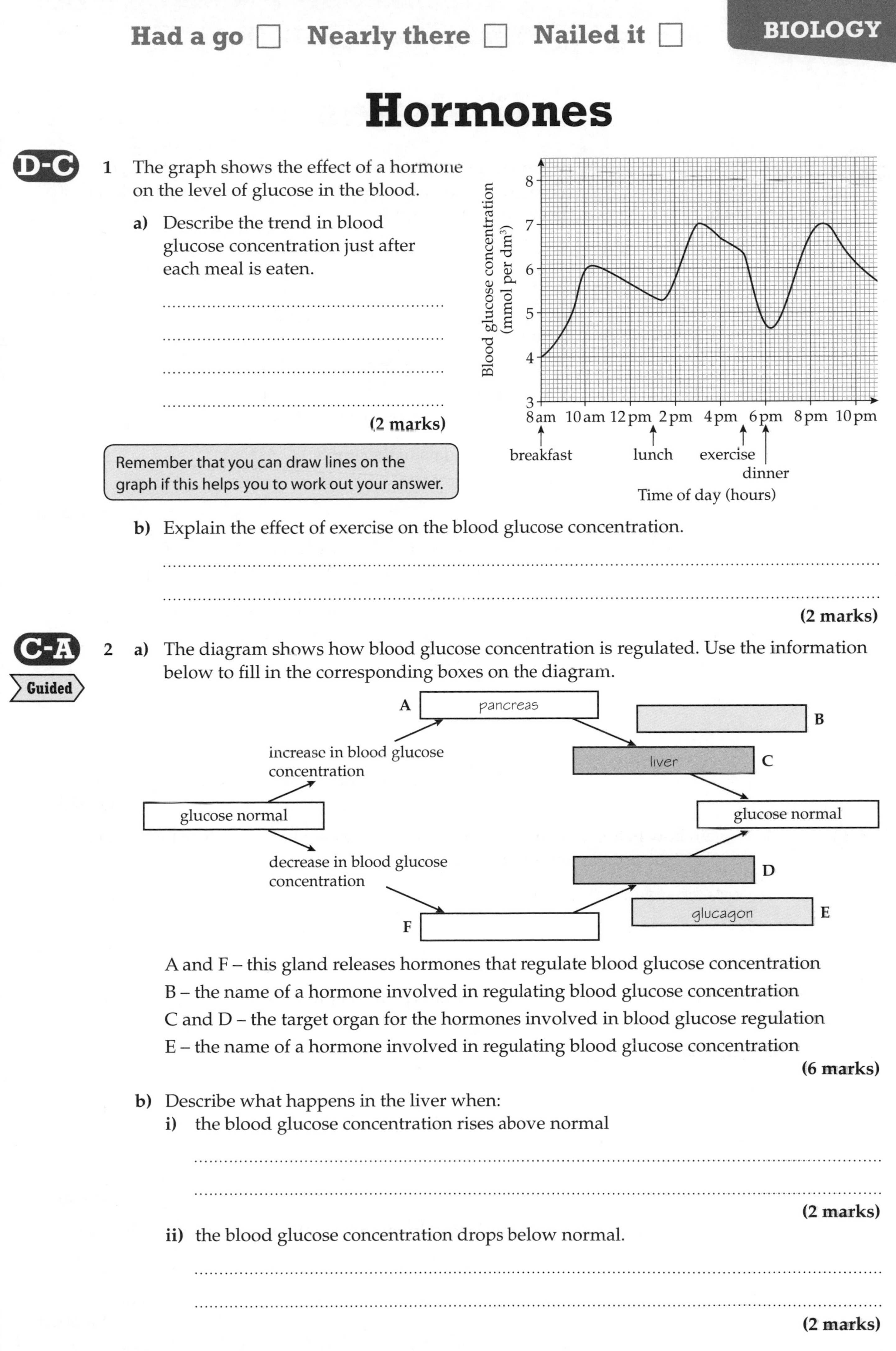

Hormones

D-C

1 The graph shows the effect of a hormone on the level of glucose in the blood.

a) Describe the trend in blood glucose concentration just after each meal is eaten.

....

....

....

....

(2 marks)

Remember that you can draw lines on the graph if this helps you to work out your answer.

b) Explain the effect of exercise on the blood glucose concentration.

....

....

(2 marks)

C-A **Guided**

2 a) The diagram shows how blood glucose concentration is regulated. Use the information below to fill in the corresponding boxes on the diagram.

A and F – this gland releases hormones that regulate blood glucose concentration

B – the name of a hormone involved in regulating blood glucose concentration

C and D – the target organ for the hormones involved in blood glucose regulation

E – the name of a hormone involved in regulating blood glucose concentration

(6 marks)

b) Describe what happens in the liver when:

i) the blood glucose concentration rises above normal

....

....

(2 marks)

ii) the blood glucose concentration drops below normal.

....

....

(2 marks)

Diabetes

C-A

1 a) The graph shows the percentage of people in one area of America in the year 2000 who have diabetes, divided into groups according to body mass index. For example, 7% of the population have diabetes and are overweight.

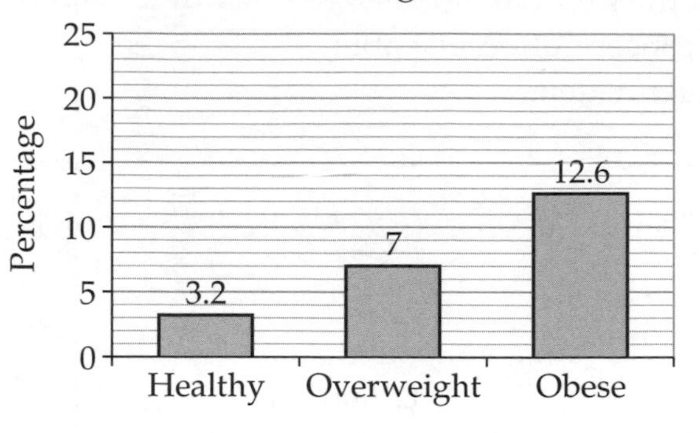

i) Describe the trend shown by the data in the graph.

....

(1 mark)

Don't forget that the axis labels are important in describing the trend of any graph. Avoid using vague statements such as 'it goes up'. Describe what is increasing by using the information given by the labels.

ii) The population size in this area of America in 2000 was 1 808 344. How many of these people suffered from diabetes?

Number of people ....

(3 marks)

Guided

b) Explain how helping people to control their diets might help to reduce the percentage of people in the population who have diabetes.

Controlling diets will help to ....

Fewer obese people means ....

(2 marks)

C-B

2 There are two types of diabetes, Type I and Type II. People who have either type of diabetes are unable to control the levels of glucose in their blood.

a) State the cause of Type 1 diabetes.

....

(1 mark)

b) State the cause of Type II diabetes.

....

(1 mark)

c) Explain why a diabetic may be able to help control their blood sugar levels with exercise.

....

....

(2 marks)

Plant hormones

D-C **1** The diagram shows how a plant shoot responds to light.

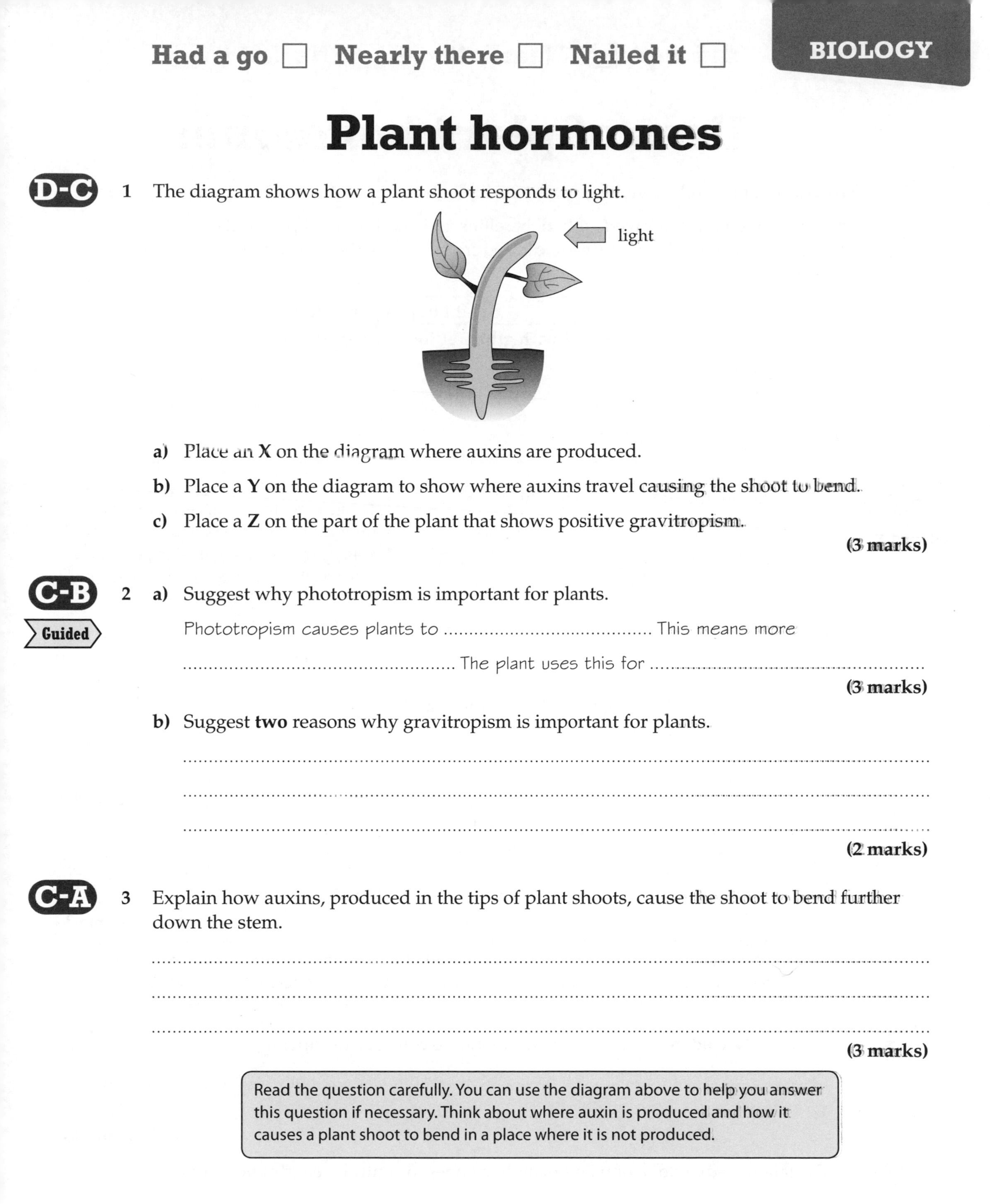

a) Place an **X** on the diagram where auxins are produced.

b) Place a **Y** on the diagram to show where auxins travel causing the shoot to bend.

c) Place a **Z** on the part of the plant that shows positive gravitropism.

(3 marks)

C-B **Guided** **2** **a)** Suggest why phototropism is important for plants.

Phototropism causes plants to .... This means more

.... The plant uses this for ....

(3 marks)

b) Suggest **two** reasons why gravitropism is important for plants.

....

....

....

(2 marks)

C-A **3** Explain how auxins, produced in the tips of plant shoots, cause the shoot to bend further down the stem.

....

....

....

(3 marks)

Read the question carefully. You can use the diagram above to help you answer this question if necessary. Think about where auxin is produced and how it causes a plant shoot to bend in a place where it is not produced.

Uses of plant hormones

The effect of gibberellin on the growth of pea plants was investigated by a group of students.

One set of plants was sprayed with gibberellins and another set of plants, the control, was sprayed with water. The plants were each 60 cm tall before the investigation. The table shows the results of the investigation.

	Height of pea plants (cm)	
	Sprayed with gibberellins	**Sprayed with water**
Plant 1	104	65
Plant 2	93	61
Plant 3	99	62
Mean		62.7

C-B

1 **a)** Calculate the mean height for the plants sprayed with gibberellins.

104+93+99 = 296

To calculate a mean you need to add all the figures together and then divide by the number of readings in your sample. In this case add the heights and divide by three.

Mean height .... cm

(2 marks)

b) Describe the effect of gibberellins on the pea plants.

....

....

(2 marks)

c) Describe the reason for using a control in this investigation.

....

....

(1 mark)

You may have used controls in the practical work that you have carried out. Think about the reasons why you used controls in these experiments. The reasons here will be the similar.

C-A

2 Describe **two** advantages of using hormones to control fruit ripening.

....

....

(2 marks)

C-A

3 Explain the advantage of using plant hormones to control weeds in fields of crops.

....

....

(2 marks)

C-A

4 Suggest why rooting powder is used by some plant growers to produce more plants.

....

....

(2 marks)

Biology extended writing 3

A class of students investigate the ability of their tongues to detect different tastes: bitter, salty, sour and sweet. The students have four solutions, each with a different taste. They also have some cotton buds.

There are 30 students in the class and they collect this data. The table shows the numbers of students who correctly identified the taste on different parts of their tongue.

Part of tongue	Bitter	Salty	Sour	Sweet
back	28	7	10	1
front	13	23	9	26
side	12	24	27	7

Suggest how the class could carry out an experiment to collect this data, and what conclusions they could draw from it.

(6 marks)

You will be more successful in extended writing questions if you plan your answer before you start writing.

The question asks you to plan a practical and to use the data. Experimental design questions may come up in the Extended Writing questions. Think about:

- how you would carry out the practical investigation, including the apparatus used
- how many results you would record, including the number and range to collect
- how to make the investigation safe and control the variables.

When looking at the conclusion you should think about:

- any clear patterns in the data
- which parts of the tongue are linked to which tastes
- the idea that some parts of the tongue may be sensitive to more than one taste.

....

....

....

....

....

....

....

....

....

....

....

....

....

....

....

Effects of drugs

C-B **1** A group of students carried out an investigation into reaction times. They each timed how long it took for them to catch a falling ruler. The results of the investigation are shown in the graph.

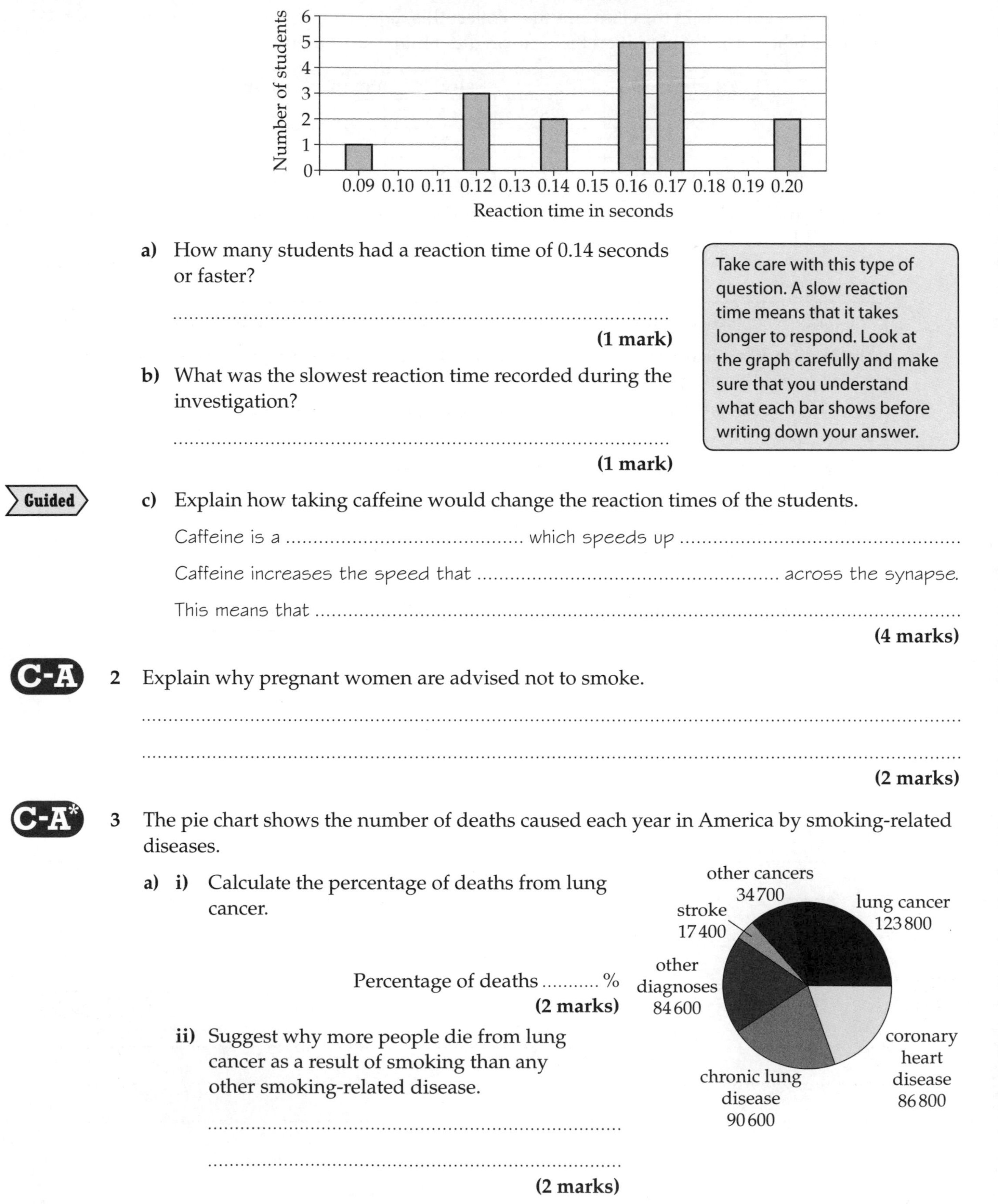

a) How many students had a reaction time of 0.14 seconds or faster?

....

(1 mark)

b) What was the slowest reaction time recorded during the investigation?

....

(1 mark)

> Take care with this type of question. A slow reaction time means that it takes longer to respond. Look at the graph carefully and make sure that you understand what each bar shows before writing down your answer.

Guided

c) Explain how taking caffeine would change the reaction times of the students.

Caffeine is a .... which speeds up ....

Caffeine increases the speed that .... across the synapse.

This means that ....

(4 marks)

C-A **2** Explain why pregnant women are advised not to smoke.

....

....

(2 marks)

C-A* **3** The pie chart shows the number of deaths caused each year in America by smoking-related diseases.

a) i) Calculate the percentage of deaths from lung cancer.

Percentage of deaths .... %

(2 marks)

ii) Suggest why more people die from lung cancer as a result of smoking than any other smoking-related disease.

....

....

(2 marks)

Effects of alcohol

D-C

1 The graph shows how the risk of having a car accident changes at different concentrations of alcohol in the blood.

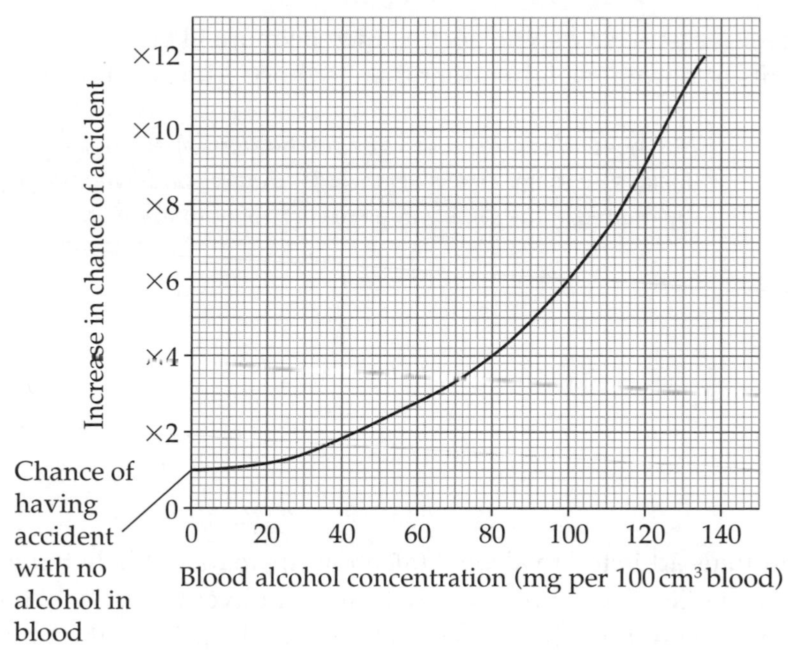

a) Describe the overall trend in the data shown by the graph.

....

....

(2 marks)

b) Drinking one glass of wine raises the blood alcohol concentration by 20 mg per 100 cm^3 of blood. Calculate the blood alcohol concentration of the blood after five glasses of wine.

Blood alcohol concentration ....

(2 marks)

c) Use the information from the graph to state the increased risk of having an accident after five glasses of wine.

....

(1 mark)

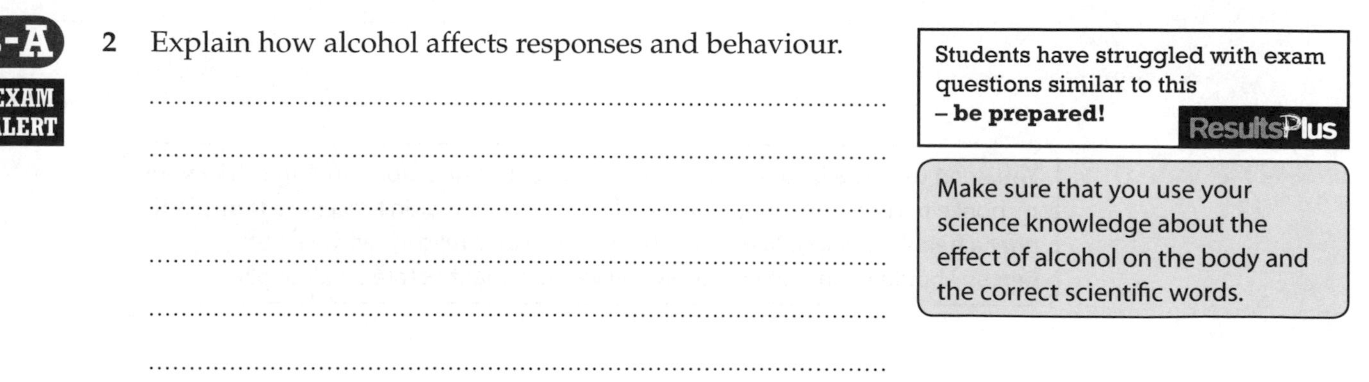

B-A

EXAM ALERT

2 Explain how alcohol affects responses and behaviour.

....

....

....

....

....

....

(4 marks)

Students have struggled with exam questions similar to this – **be prepared!** ResultsPlus

Make sure that you use your science knowledge about the effect of alcohol on the body and the correct scientific words.

Ethics and transplants

1 A study that was carried out has found that clinically obese patients are 44% less likely to receive a kidney transplant than a person of normal weight.

Suggest reasons why some doctors may be more likely to authorise kidney transplants to patients of normal weight rather than to those who are clinically obese.

> You need to consider both sides of the story when answering this question. Make sure that your answer includes details about why obese patients may not be treated in the same way as patients of normal weight.

(4 marks)

C-A

2 Individuals addicted to alcohol (alcoholics) are more likely to need a liver transplant than non-alcoholics. Some hospitals that carry out liver transplants insist that the alcoholic patient must not drink any alcohol for 6 months prior to the transplant being carried out. This is known as the 6-month rule.

a) Discuss the reasons for some hospitals insisting on the 6-month rule.

(3 marks)

> This question is not asking for your opinion so you do not need to present an argument about whether the alcoholic *should* receive a transplant. You do need to think carefully about why the 6-month rule was made in the first place and how this might affect the success of the operation.

b) Many people, whether they drink alcohol or not, require liver transplants and there is a shortage of donor organs.

Discuss the ethical reasons why some hospitals may carry out transplants on people who do not drink alcohol before they carry out transplants on alcoholics.

(3 marks)

> You need to give a balanced opinion to answer this question. Do not restrict your response to discussing why an alcoholic should or shouldn't receive a transplant *after* a healthy person. Make sure that you also give reasons why a healthy person should or should not receive a liver transplant before an alcoholic.

Pathogens and infection

1 The table shows the percentage of 15 to 49 year olds with HIV in some African countries.

	% of 15 to 49 year olds with HIV in some African countries			
African country	**2006**	**2007**	**2008**	**2009**
Namibia	15	14.3	13.7	13.1
South Africa	18.1	18	17.9	17.8
Zambia	13.8	13.7	13.6	13.5
Zimbabwe	17.2	16.1	15.1	14.3

D-C

a) Calculate the largest decrease in the percentage of HIV between 2008 and 2009 and state which country the decrease applied to. Use the data in the table to support your answer.

First work out what the percentage decrease was for each county. Some of them are easy. For example Zambia went from 13.6% to 13.5%. If you are not sure – use your calculator!

....

....

(2 marks)

b) The data for each African country follows the same overall trend. Use the data in the table to describe this trend.

....

....

(2 marks)

D-C

2 Suggest why dentists are required to wear gloves when working on the teeth of their patients.

....

(1 mark)

C-B Guided

3 State **two** diseases caused by pathogens and describe how these diseases are transmitted from one person to another.

As part of the guidance the first one has been done for you.

Name of disease: Cholera

How disease is transmitted: Cholera is spread through dirty water.

Name of disease: ....

How disease is transmitted: ....

(4 marks)

C-A

4 Describe the role of the skin in protecting the body from disease.

....

....

(2 marks)

B-A

5 Some substances in cigarette smoke prevent cilia in the airways from functioning normally. Explain how this may lead to an increase in the risk of lung infections in smokers.

....

....

....

(3 marks)

Antiseptics and antibiotics

1 Complete the table to show the type of substance you would use to safely destroy microorganisms in each of the situations given.

There are lots of different key words here. Make sure you understand the differences between all the substances used to kill microorganisms. In the table, one of the answers has been given as part of the guided text.

Situation	Substance used to safely destroy microorganism
An open wound in the skin	
Athlete's foot	Anti-fungal cream
Removing bacteria from a kitchen surface	
A cholera infection	

(4 marks)

D-C

2 The graph shows the results of an investigation into ways of killing bacteria in the mouth. The investigation looked at the effectiveness of two different types of fruit juice.

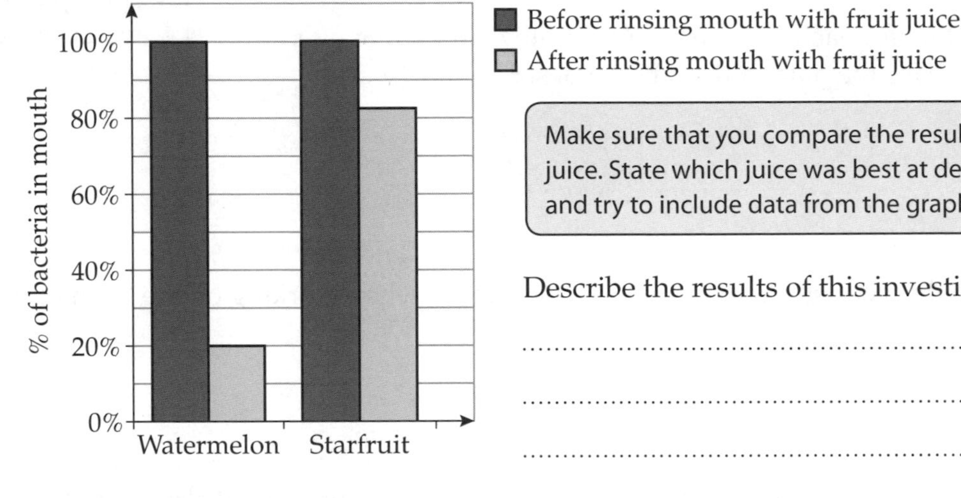

Make sure that you compare the results for each type of juice. State which juice was best at destroying the bacteria and try to include data from the graph in your answer.

Describe the results of this investigation.

....

....

....

....

(2 marks)

B-A

3 Vancomycin is the most common and the most effective antibiotic used in the treatment of MRSA infections. A new strain of MRSA is beginning to develop a resistance to vancomycin.

a) Suggest why vancomycin-resistant MRSA is causing concern among doctors and scientists.

....

....

(2 marks)

b) Suggest how vancomycin-resistant MRSA have developed from non-resistant forms of the bacteria.

....

....

(2 marks)

c) Suggest why scientists are investigating the use of plants to help in the fight against vancomycin-resistant MRSA.

....

....

(2 marks)

Interdependence and food webs

D-C

1 The diagram shows a food web.

Explain what will happen to the number of mice if the owls are removed from the food web.

....

....

....

....

....

....

(3 marks)

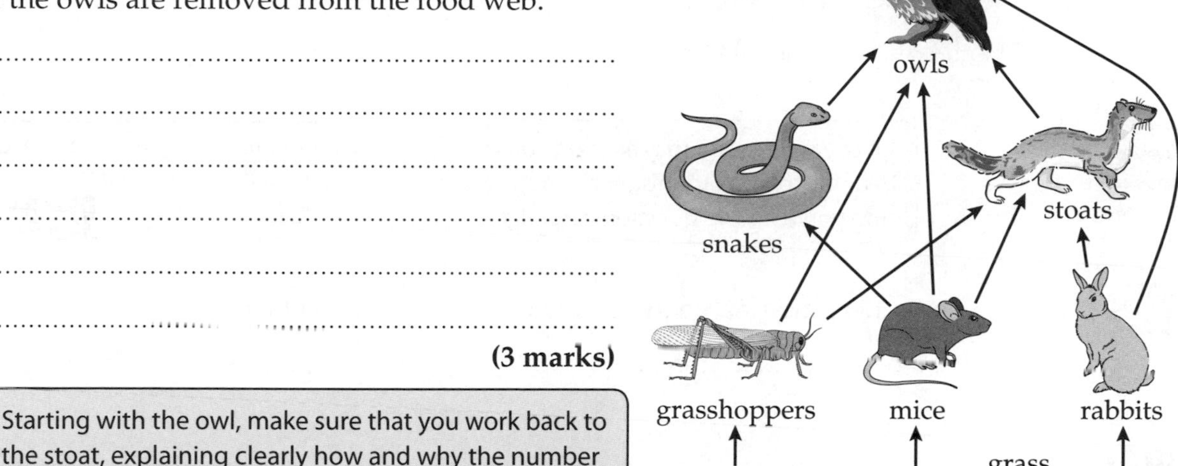

Starting with the owl, make sure that you work back to the stoat, explaining clearly how and why the number of mice are affected by the removal of the owl.

C-A

Guided

2 a) The table gives details about the trophic levels in a food chain.

Calculate the biomass at each trophic level. Write your answers in the spaces in the table.

Organism	Energy at each trophic level (J)	Number of organisms	Mass of each organism (kg)	Biomass at each trophic level (kg)
Producers	7550	10 000	0.25	2500
Herbivores	640	200	2.5	
Carnivores	53	10	20	

(3 marks)

EXAM ALERT

b) Use your calculations to draw a pyramid of biomass for the food chain shown in the table.

If you are asked to draw a pyramid of biomass, make sure you draw it to scale.

Students have struggled with exam questions similar to this – **be prepared!** ResultsPlus

(2 marks)

c) Calculate the percentage of energy that is transferred from the herbivores to the carnivores.

Percentage of energy ....

(2 marks)

d) Explain why the amount of energy decreases as it is transferred from one trophic level to the next.

....

....

(2 marks)

Parasites and mutualists

D-C **1** Cleaner fish live on larger fish such as sharks. Suggest how the cleaner fish and the larger fish benefit from a mutualistic relationship.

Guided

Cleaner fish get food by ....

This helps the larger fish because ....

(2 marks)

EXAM ALERT

An exam question may ask you about the benefits to one organism or to both. Make sure you read the question carefully!

Students have struggled with exam questions similar to this – **be prepared!** ResultsPlus

D-C **2** State why fleas are unable to survive without an animal host.

....

(1 mark)

D-C **3** The scabies mite is a tiny insect that burrows into human skin and lays its eggs. Infection by the scabies mite causes severe itching and a lumpy, red rash that can appear anywhere on the body. Suggest why the scabies mite is a parasite and not a mutualist.

....

....

(2 marks)

This type of question is expecting you to apply your understanding of science to a situation that you may not be familiar with. You will have been taught about organisms that behave in a similar way to the scabies mite – use what you know about these organisms but apply it to the scabies mite.

C-A **4** Explain how the relationship between chemosynthetic bacteria and tube worms in deep sea vents benefits each organism.

....

....

....

(3 marks)

B-A **5** Describe how legume plants such as peas benefit from a mutualistic relationship with bacteria.

....

....

....

(3 marks)

Pollution

D-C

1 a) Name two pollutants that can be released into the environment as a result of the overuse of fertilisers.

....

....

(2 marks)

b) Describe **two** ways that an increase in the human population can lead to an increase in atmospheric pollution.

....

....

....

....

(4 marks)

C-A

2 The graph shows the mass of fertiliser used in the world from 1950 to 2003.

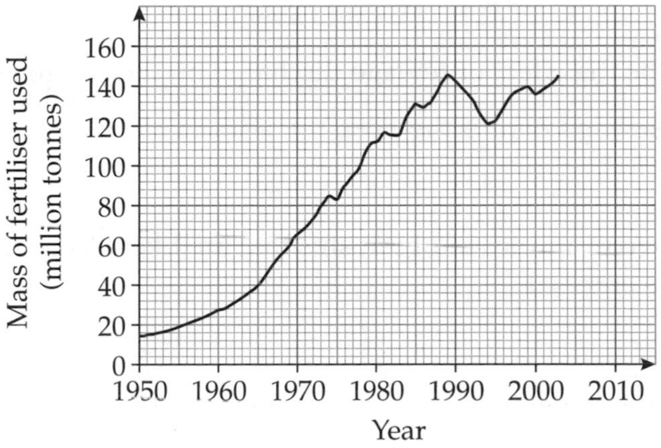

Mass of fertiliser used (million tonnes)

0, 20, 40, 60, 80, 100, 120, 140, 160

1950 1960 1970 1980 1990 2000 2010

Year

Guided

a) Calculate the percentage increase in fertiliser use from 1950 to 2003.

145 − 15 = ...

(......................../15) × 100 =

Percentage increase ....

Make sure that you read the graph carefully to get the correct figures for your calculation. You will get one mark for showing the correct calculation and one mark for the correct answer.

(2 marks)

b) Suggest why there has been a change in the mass of fertiliser used worldwide since 1950.

....

....

....

(2 marks)

c) Suggest why the mass of fertiliser that farmers can use on their crops is regulated.

....

....

(2 marks)

Pollution indicators

D-C **1** Suggest **two** reasons why people are encouraged to recycle materials.

....

....

(2 marks)

2 The graph shows the oxygen concentration in a river. At one point sewage enters the river. Sewage contains high levels of nitrates.

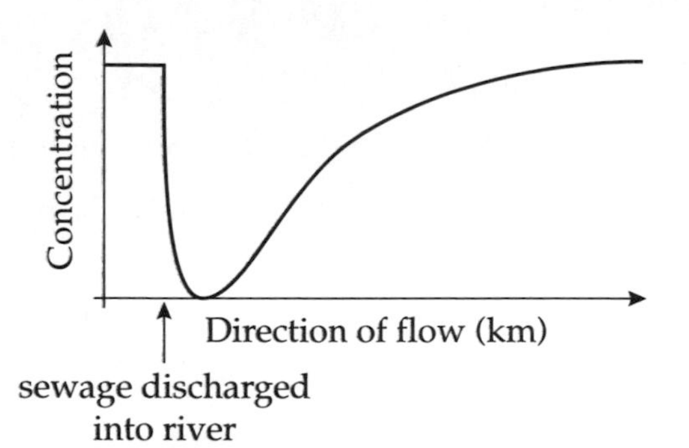

Guided

a) Describe the effect of sewage discharge on the oxygen content of the river.

When sewage enters the river, the amount of oxygen .... and

then .... **(2 marks)**

b) Give two examples of organisms that are likely to be found in the river before the sewage enters it.

....

....

(2 marks)

c) Sewage contains nitrates. Describe the effect that the nitrates will have on living organisms in the river.

....

....

....

....

(4 marks)

You need to think through the sequence of events that occurs after eutrophication very carefully. Describe these events in the order in which they happen.

C-A **3** Describe how **two** different organisms can be used to help monitor air quality.

....

....

....

....

(4 marks)

The carbon cycle

D-C Guided

1 Complete the diagram of the carbon cycle by writing the names of the processes in the boxes.

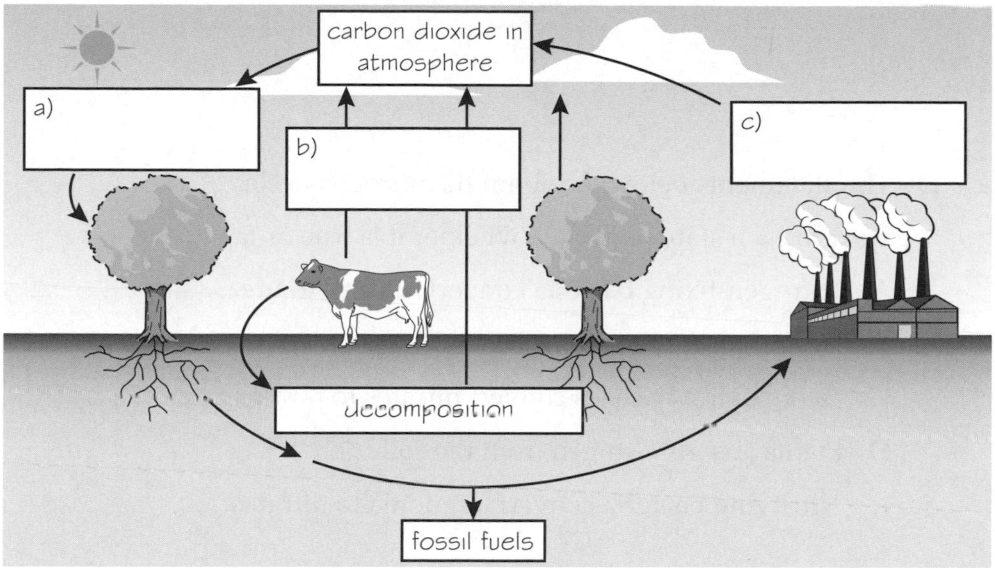

carbon dioxide in atmosphere

a)

b)

c)

decomposition

fossil fuels

(6 marks)

D-B

2 Explain why bacteria are important in recycling carbon in the environment.

....

....

(2 marks)

EXAM ALERT

Students have struggled with exam questions similar to this – **be prepared!** ResultsPlus

In questions about the carbon cycle, you will be expected to make links between photosynthesis, respiration and combustion, and the amount of carbon dioxide in the air.

C-A

3 **a)** The diagram shows a fish tank. Explain how carbon is recycled between organisms in the fish tank.

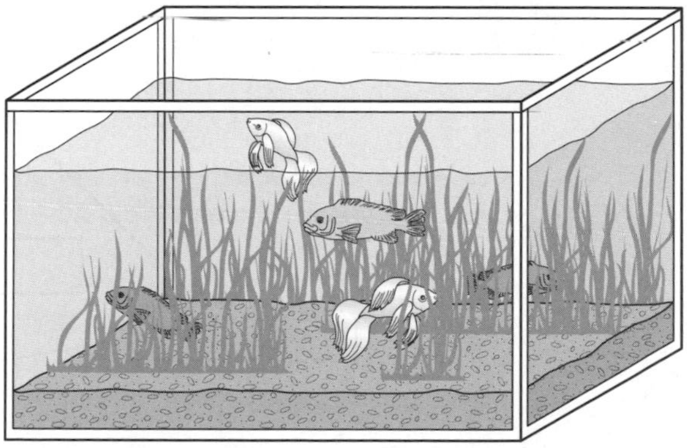

....

....

....

....

....

....

(4 marks)

b) Suggest why it is important that both plant and animal populations in the fish tank are kept balanced.

....

....

....

(3 marks)

The nitrogen cycle

D-C **1** Explain why nitrogen compounds are important for plants.

....

....

(2 marks)

C-B **2** **a)** The statements below are about the nitrogen cycle.

Next to each statement state whether it is true or false.

A Nitrogen-fixing bacteria convert nitrates to urea. ....

B Decomposers live in the root nodules of legumes. ....

C Denitrifying bacteria convert nitrates to nitrogen gas. ....

D Plants absorb nitrogen from the soil. ....

E Nitrifying bacteria convert ammonia to nitrates. ....

(5 marks)

b) Describe the role of decomposers in the nitrogen cycle.

....

....

(2 marks)

C-A **3** The diagram shows the nitrogen cycle.

One of the boxes has been filled in for you as part of the guided question.

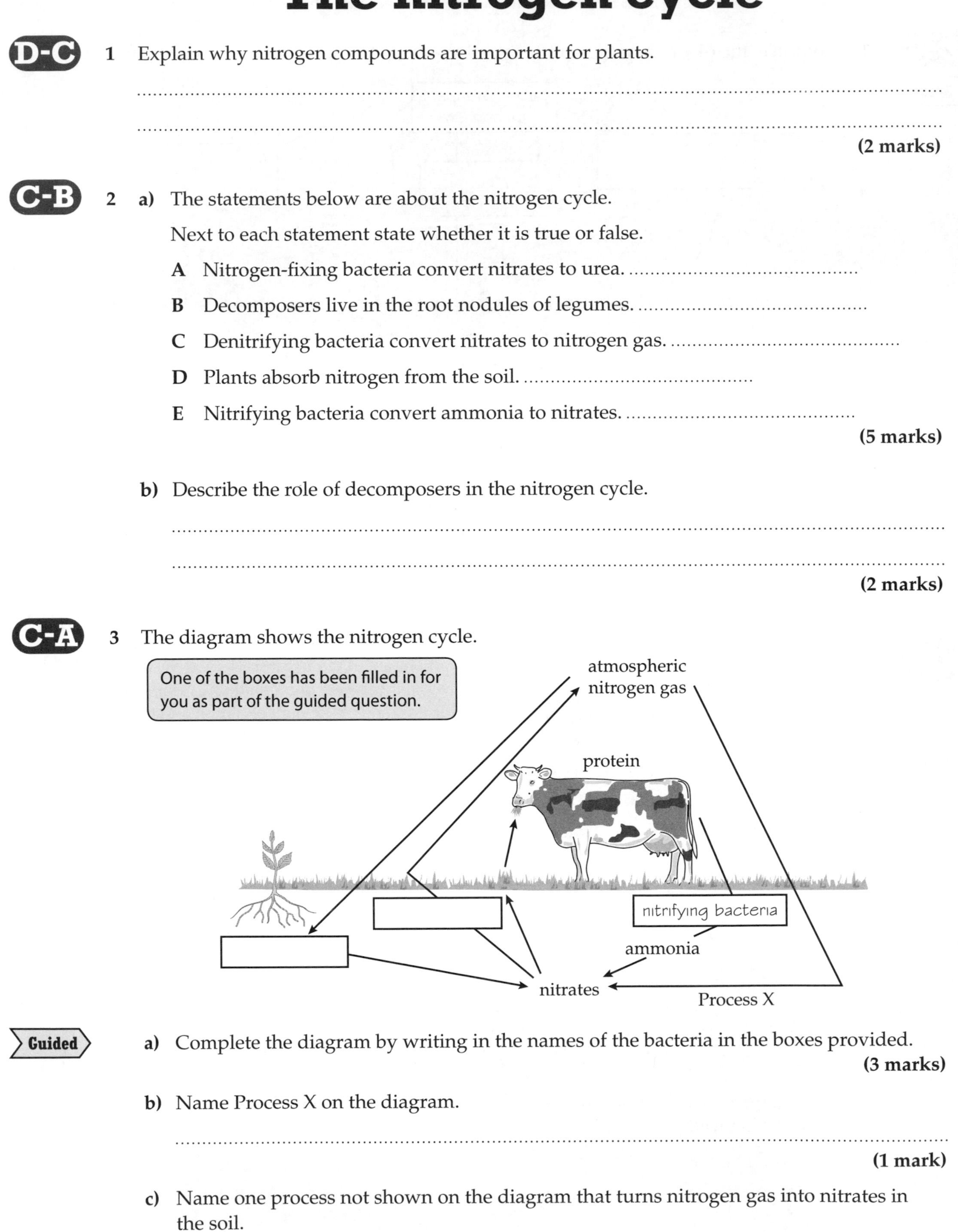

Guided **a)** Complete the diagram by writing in the names of the bacteria in the boxes provided.

(3 marks)

b) Name Process X on the diagram.

....

(1 mark)

c) Name one process not shown on the diagram that turns nitrogen gas into nitrates in the soil.

....

(1 mark)

Biology extended writing 4

Bronchitis is a disease of the lungs. People with bronchitis have a persistent cough. The disease is often caused by either smoking or air pollution.

Scientists at Edinburgh University did a study on people between the ages of 35 and 69 years old, to find the percentage who had bronchitis. They looked at smokers and non-smokers, living in areas of high and low atmospheric pollution.

The graph shows their results.

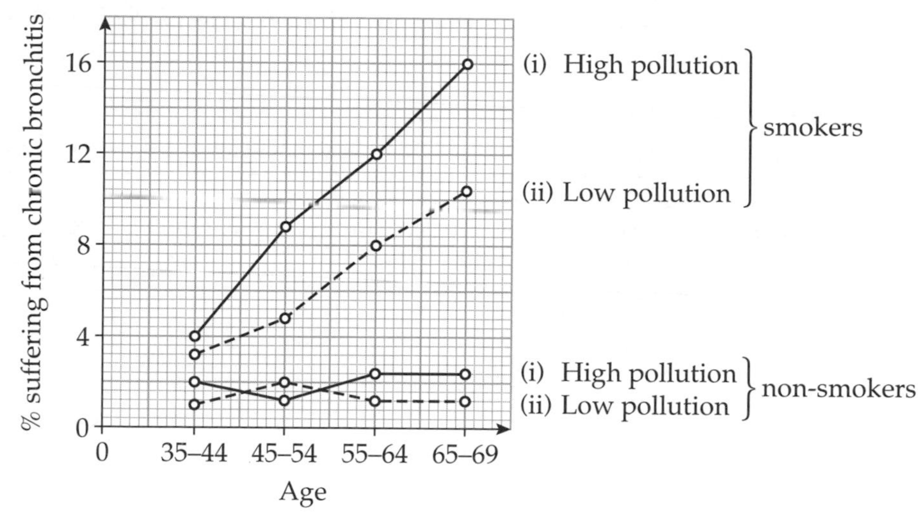

Explain what the data shows about the relative effects of smoking and atmospheric pollution on the incidence of bronchitis.

(6 marks)

> You will be more successful in extended writing questions if you plan your answer before you start writing.
>
> The question asks you to interpret the graph. Spend some time to look at the information. Think about the following things:
>
> - Identify the data for smokers and non-smokers. How do these two groups relate to each other?
> - What do the dotted lines and the continuous lines show?
> - How do the rates of bronchitis change in older people in the study?
>
> It will help your answer if you use some data from the graph to support your answer.

....

....

....

....

....

....

....

....

....

....

....

Biology extended writing 5

Some organisms exist in a feeding relationship called mutualism.

Describe what is meant by mutualism, showing how this relationship works with some example organisms.

(6 marks)

You will be more successful in extended writing questions if you plan your answer before you start writing.

The question asks you to describe what mutualism is and give some examples. Think about:

- how you would define mutualism
- how a mutalistic relationship works
- some examples of mutual relationships including how the relationship works for both organisms.

Don't forget to use as many examples as you can – there are four mutualistic relationships on the specification. Can you remember them all?

The early atmosphere

E-C **1** According to a model used by some scientists, the earliest atmosphere consisted of the various gases shown in the pie chart.

10%
80%

Key
carbon dioxide
water vapour
other gases

a) State the percentage of carbon dioxide that should go in this pie chart.

.... **(1 mark)**

b) Scientists have found many different sources of information about the early atmosphere. Suggest why this information does not help us to know clearly what the early atmosphere was like.

....

.... **(1 mark)**

B-A* 

2 Some scientists think that the early atmosphere of the Earth may have been similar to that of the planet Venus. The pie chart on the right shows the gases in the atmosphere of Venus.

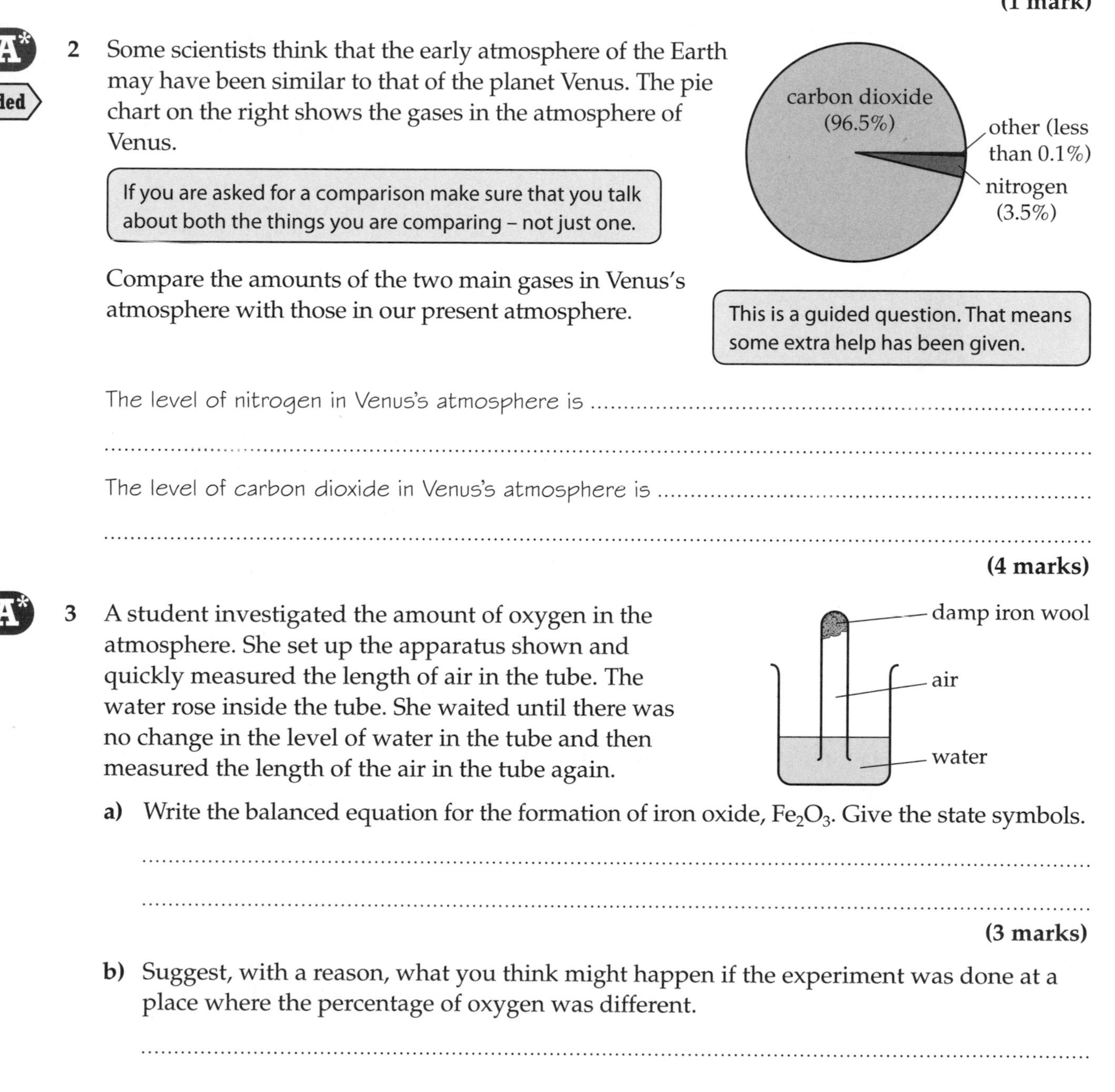

If you are asked for a comparison make sure that you talk about both the things you are comparing – not just one.

Compare the amounts of the two main gases in Venus's atmosphere with those in our present atmosphere.

This is a guided question. That means some extra help has been given.

The level of nitrogen in Venus's atmosphere is ....

....

The level of carbon dioxide in Venus's atmosphere is ....

.... **(4 marks)**

B-A* **3** A student investigated the amount of oxygen in the atmosphere. She set up the apparatus shown and quickly measured the length of air in the tube. The water rose inside the tube. She waited until there was no change in the level of water in the tube and then measured the length of the air in the tube again.

a) Write the balanced equation for the formation of iron oxide, Fe_2O_3. Give the state symbols.

....

.... **(3 marks)**

b) Suggest, with a reason, what you think might happen if the experiment was done at a place where the percentage of oxygen was different.

....

.... **(2 marks)**

A changing atmosphere

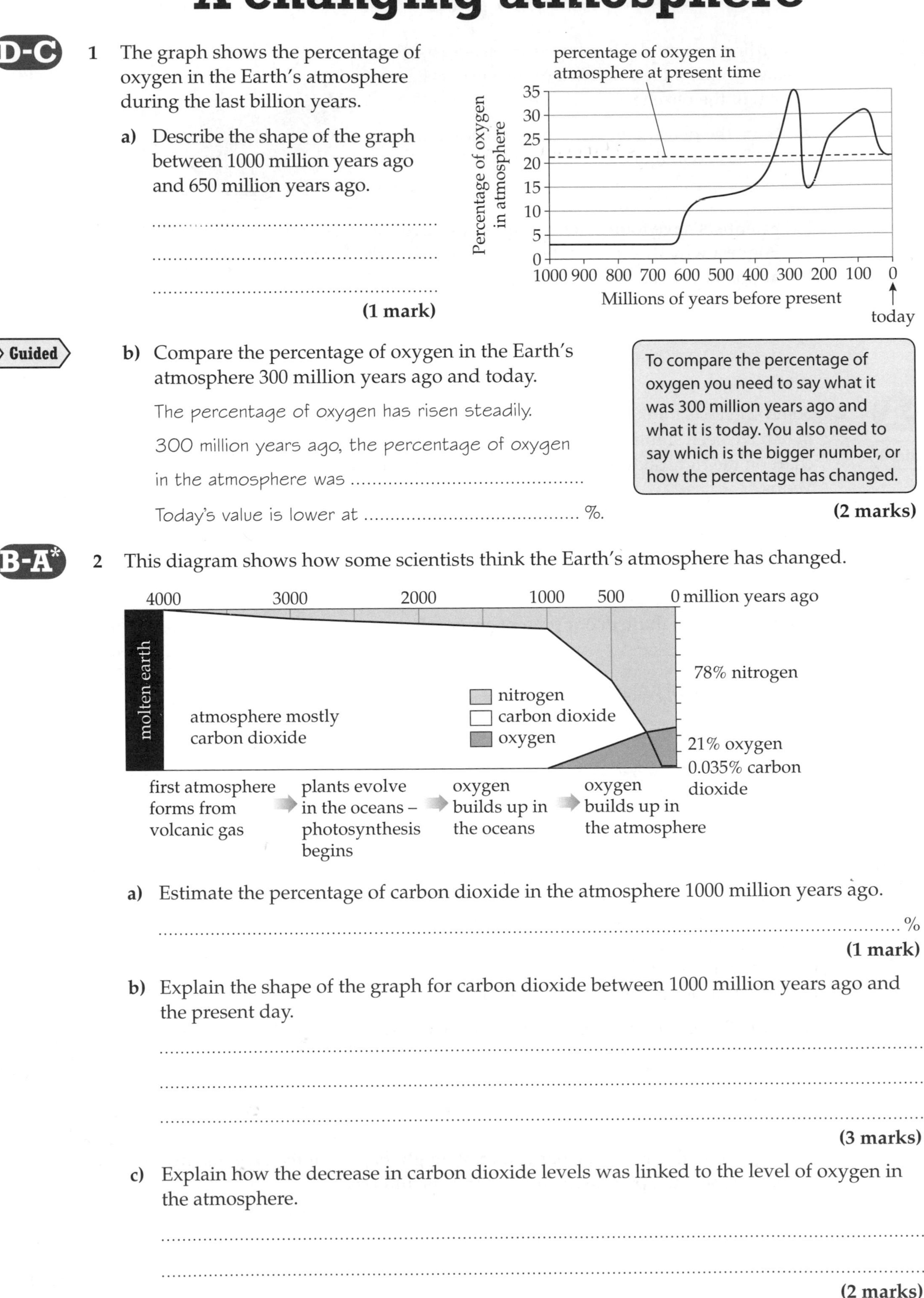

D-C 1 The graph shows the percentage of oxygen in the Earth's atmosphere during the last billion years.

a) Describe the shape of the graph between 1000 million years ago and 650 million years ago.

....

....

....

(1 mark)

Guided **b)** Compare the percentage of oxygen in the Earth's atmosphere 300 million years ago and today.

The percentage of oxygen has risen steadily.

300 million years ago, the percentage of oxygen in the atmosphere was ....

Today's value is lower at .... %.

> To compare the percentage of oxygen you need to say what it was 300 million years ago and what it is today. You also need to say which is the bigger number, or how the percentage has changed.

(2 marks)

B-A* 2 This diagram shows how some scientists think the Earth's atmosphere has changed.

a) Estimate the percentage of carbon dioxide in the atmosphere 1000 million years ago.

.... %

(1 mark)

b) Explain the shape of the graph for carbon dioxide between 1000 million years ago and the present day.

....

....

....

(3 marks)

c) Explain how the decrease in carbon dioxide levels was linked to the level of oxygen in the atmosphere.

....

....

(2 marks)

Rocks and their formation

1 Use the words from the box to fill in the gaps in the following sentences:

a)

igneous	metamorphic	sedimentary	man-made

Limestone is a s.edimentary....... rock. Marble is a m.etamorphic....... rock.

(2 marks)

b) State the link between calcium carbonate, limestone and marble.

limestone and marble are both made of calcium carbonate.

(1 mark)

D-C

2 Explain why sedimentary rock may contain fossils but igneous rock will not.

Because sedimentary rock are formed under the sea and the fossil are trapped under the sediment, but igneous rock is formed from lava so the heat would cause the fossils to

(3 marks)

B-A*

EXAM ALERT

3 Explain how metamorphic rock may be formed from sedimentary rock.

When heat and pressure is applied to the sedimentary rock new crystals are formed and they become metamorphic rocks

from layers above it

(2 marks)

Students have struggled with exam questions similar to this – **be prepared!** ResultsPlus

You need to revise how different rocks are formed.

B-A*

4 a) Describe how granite is formed and what it looks like.

Granite is formed when magma cools slowly, it is made up of large ~~crys~~ crystal which you can see in the rock.

(silidapys)

(3 marks)

The question asks you for two things – how granite forms and what it looks like – don't forget to answer both parts of the question.

b) Explain why igneous rocks that were formed deep underground often contain different crystals to igneous rocks formed closer to the surface.

Igneous rocks that formed under ground cools slower so therefore the crystals are larger. Igneous rocks that are formed closer to the surface cool faster and the rock are therefore smaller.

(2 marks)

Limestone and its uses

1 State two advantages and two disadvantages of quarrying limestone.

Two advantages are that it provides jobs, so is good for the ~~envir~~ economy and it is used to make building materials such as cement ~~concrete~~. Two disadvantages are that is creates noice and air pollution and damages habitats

(4 marks)

Think about economic, social and environmental factors when thinking about advantages and disadvantages.

B-A* 2 The diagram shows how cement is made.

Name the type of chemical reaction that the limestone undergoes when it passes through the rotating kiln and write a balanced equation for the reaction involved.

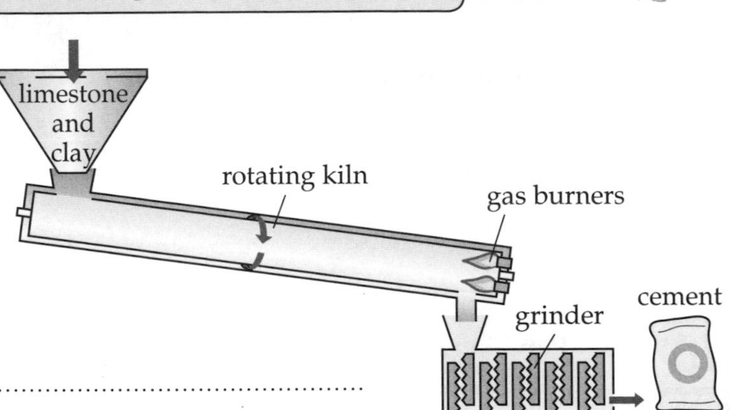

....

....

(3 marks)

B-A* 3 The table shows the main uses for the limestone mined at a quarry.

EXAM ALERT

Use	Percentage (%)	Number of tonnes used
construction	60	210000
making cement	20	70000
making steel and iron alloys	5	17 500
other uses	15	52500

Students have struggled with exam questions similar to this – **be prepared!** ResultsPlus

Some questions in the exam may ask you to read data from a table. Make sure that you use only the relevant data from the table.

Guided

a) Complete the table to show the number of tonnes for making cement and for other uses.

5% is the same as 17 500 tonnes

20% is four times as much so 70000 tonnes

15% is 3 times as much so 52500 tonnes

60% is 12 times as much so 210000 tonnes

(3 marks)

b) Calculate the total output from the quarry.

70000
52500
210000
332500

Total output 332500 tonnes

(2 marks)

c) Suggest, with a reason, how an increase in the number of houses being built will affect the price of cement.

....

....

(2 marks)

Formulae and equations

1 In the table place two ticks (✓) in each row for each substance given.

Substance	Atom	Molecule	Element	Compound
H_2O		✓		✓
O_2		✓	✓	
Ca	✓		✓	

4

(3 marks)

It is important to learn these definitions and other key scientific words. Make sure you understand the meanings of words such as atom, molecule, element, mixture and compound.

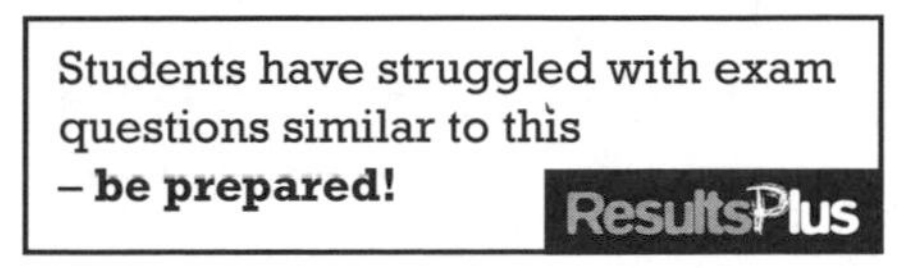

Students have struggled with exam questions similar to this – **be prepared!** ResultsPlus

D-C Guided

2 In each large circle draw the atoms present in each of these materials. Use the key to show the different types of atom.

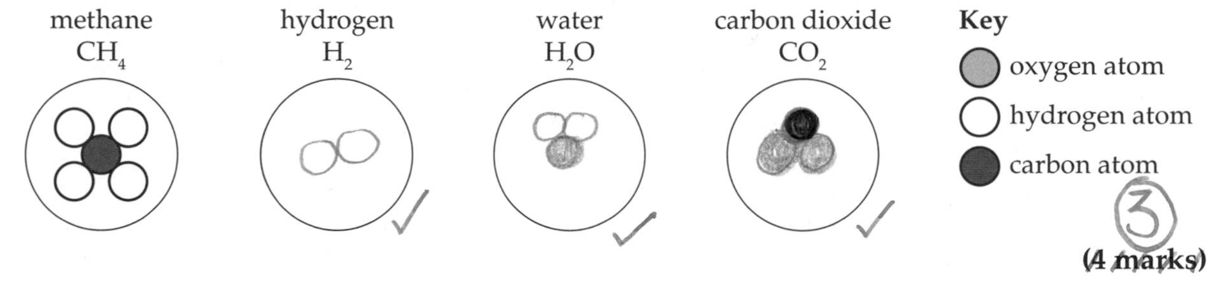

3

(4 marks)

B-A*

3 The balanced equation for the reaction of carbon dioxide with calcium hydroxide, $Ca(OH)_2$ also known as lime water, is:

$$Ca(OH)_2(\ldots\ldots) + CO_2(\ldots\ldots) \rightarrow CaCO_3(\ldots\ldots) + H_2O(\ldots\ldots)$$

a) Explain what the number 2 in $Ca(OH)_2$ means.

The 2 only applies for the part in the bracket, so in this case there are 2 oxygen and hydrogen atoms.

2

(2 marks)

b) Write the state symbols in the brackets after each substance in the equation.

$1\frac{1}{2}$

(2 marks)

B-A*

4 Sodium hydrogen carbonate, $NaHCO_3$, decomposes into sodium carbonate Na_2CO_3, carbon dioxide and water.

$$2NaHCO_3\ (s) \rightarrow Na_2CO_3\ (s) + CO_2\ (g) + H_2O\ (l)$$

Explain the difference in the use of the number 2 in each of these: 2Na … and Na_2 …

2Na means that there are 2 sodium carbonate molecules.

Na_2 means the each molecule contains two sodium atoms

2

(2 marks)

$12\frac{1}{2}$ / 13

Chemical reactions

D-C EXAM ALERT

1 a) Silver chloride and sodium nitrate are formed when silver nitrate and sodium chloride solutions are mixed together. Write the word equation for this reaction.

....

....

....

(2 marks)

b) All these substances are soluble in water except silver chloride. Give the name of this reaction.

....

(1 mark)

Students have struggled with exam questions similar to this **– be prepared!** ResultsPlus

If you are asked to write a word equation, do not be tempted to use formulae (such as CO_2) as a shortcut. If you put the formulae in instead of the names, you will only get the marks for a question if the equation is correctly balanced. Stick to the words!

E-C

2 6 g of lead nitrate solution is mixed with 6 g of potassium iodide solution. State the total mass of the mixture after the reaction and explain your answer.

....

....

(2 marks)

D-C

3 The reaction that occurs when magnesium carbonate is heated is:

magnesium carbonate → magnesium oxide + carbon dioxide

In a chemical reaction, the total number of atoms is the same before and after the reaction. This means the total mass is the same before and after.

When 4.2 g of magnesium carbonate were heated until the reaction was complete, only 2.0 g of magnesium oxide was left. What mass of carbon dioxide was produced?

Mass of carbon dioxide .... g

(1 mark)

D-C Guided

4 The table shows the name, formula and atomic content of some molecules. Complete the table.

Name	Formula	Sodium atoms	Hydrogen atoms	Nitrogen atoms	Sulfur atoms	Oxygen atoms
sodium nitrate	$NaNO_3$	1		1		3
nitric acid	HNO_3					
sodium hydroxide	NaOH					
sodium sulfate	Na_2SO_4					
sulfuric acid	H_2SO_4					

(5 marks)

D-B

5 The formula for calcium hydroxide is $Ca(OH)_2$. Calcium hydroxide reacts with nitric acid, HNO_3, to form calcium nitrate and water. The formula for calcium nitrate is $Ca(NO_3)_2$.

Balance this equation for the reaction between calcium hydroxide and nitric acid.

$Ca(OH)_2$ (aq) + ... HNO_3 (aq) → $Ca(NO_3)_2$ (aq) + ... H_2O (l)

(2 marks)

Reactions of calcium compounds

E-C

1 Calcium carbonate is sprayed into the chimneys of power stations that are burning fossil fuels. Explain why this is done.

....

....

(2 marks)

The chart shows four reactions starting with heating some calcium carbonate. Sometimes reactants are added to the reactions and sometimes products are given off.

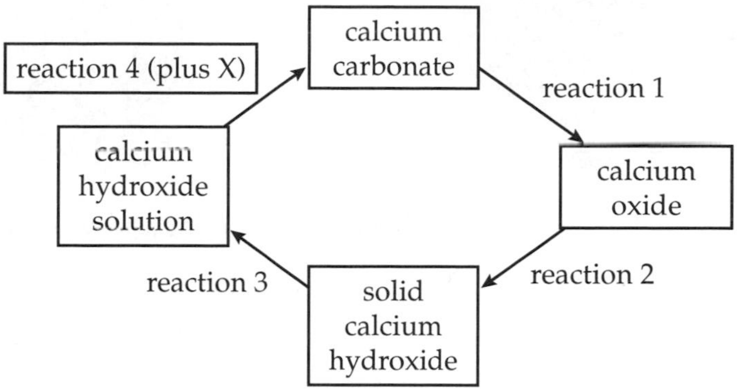

calcium carbonate → (reaction 1) → calcium oxide → (reaction 2) → solid calcium hydroxide → (reaction 3) → calcium hydroxide solution → reaction 4 (plus X) → calcium carbonate

D-C

2 State the name for reaction 1.

....

(1 mark)

D-C

3 A substance is added to calcium oxide to make reaction 2 happen. Name the substance that is added in reaction 2.

....

(1 mark)

D-C Guided

4 A gas is added to calcium hydroxide solution to make reaction 4 happen. Explain why scientists find reaction 4 useful when testing gases.

When .... is added to calcium hydroxide it turns the solution milky.

This is the test ....

(2 marks)

D-B

5 State what you would see during reaction 2.

....

....

(2 marks)

EXAM ALERT

Students have struggled with exam questions similar to this – **be prepared!** ResultsPlus

Sometimes an exam question may ask you to describe what you see when a reaction happens. Make sure you only write about what you can see.

B-A*

6 Write the balanced equation for reaction 1.

....

....

(2 marks)

Chemistry extended writing 1

A student is collecting some rocks that he knows have been formed in volcanoes. He collects two rock samples: rock A and rock B. He found rock A near a volcano that erupted 10 years ago. He found rock B near an ancient volcano that had been eroded. The erosion exposed rock that had never erupted but had cooled from magma deep underground.

This is A

This is B

Explain how the rocks that he has collected are similar and why they also show some differences in their structure.

(6 marks)

You will be more successful in extended writing questions if you plan your answer before you start writing.

In this question, you need to think about how rocks produced from a volcano are the same, and how they are different. Some questions you might like to think about are:

- What type of rocks are these?
- How do they form?
- What sort of properties would you see in this type of rock?
- How will this sort of rock be formed differently in the places where it was found?

When you are asked to explain something, you will often need to state the information and then give reasons for it.

Both rocks are igneous rocks, which means that they form from magma/lava from volcanoes, which is also known as molton rock. The rock forms when lava cools down. Igneous rocks contain hard crystals; when the lava cools ~~quickly~~ slowly the crystals are larger, but when the lava cools quickly the crystal that are formed are smaller. ~~This is because~~ The magma that cools inside the volcano are the rocks with the big ~~rocks~~ crystals because they take longer to cool, whereas the magma that cools outside the volcano form smaller crystals because they cool faster.

5/6

Use keywords intrusive/extrusive

Indigestion

1 a) State two benefits of having hydrochloric acid in our stomach.

....

....

(2 marks)

b) One substance that is used to neutralise hydrochloric acid in the stomach is magnesium hydroxide. Write the word equation for this reaction.

....

....

(2 marks)

2 The table shows some information about four indigestion tablets – O, P, R and T.

Tablet	Mass of calcium carbonate in one tablet /mg	Maximum number of tablets per day	Cost per tablet /p	Cost per 100 mg of calcium carbonate /p
O	500	16	3.5	
P	500	16	8.1	1.6
R	625	12	22.5	4.5
T	680	10	6.1	0.9

Guided

a) Calculate the cost per 100 mg of calcium carbonate in tablet O.

Cost of 1 tablet of O is ....

This contains 500 mg, so cost of 100 mg = ... / ....

Cost per 100 mg = ... p

(2 marks)

b) The ingredient in the tablets that helps to cure the indigestion is calcium carbonate. Put a cross (☒) in the box next to the statement that is correct.

☐ **A** The maximum number of tablets allowed per day rises as the cost per tablet rises.

☐ **B** The maximum number of tablets allowed per day falls as the cost per 100 mg rises.

☐ **C** The maximum number of tablets allowed per day rises as the mass of calcium carbonate in them falls.

☐ **D** The mass of calcium carbonate rises as the cost per tablet rises.

(1 mark)

c) Explain which tablet is the best value for money.

....

....

(2 marks)

3 Write the balanced equation for the reaction between calcium carbonate and hydrochloric acid.

There are some common formulae that it is helpful to learn. For example, calcium carbonate is $CaCO_3$, calcium chloride is $CaCl_2$ and hydrochloric acid is HCl.

....

....

....

(2 marks)

Neutralisation

E-C

1 Complete these word equations.

a) magnesium oxide + nitric acid → magnesium nitrate + Water

b) sulfuric acid + calcium carbonate → Calcium sulfate + water + Carbon dioxide **(2 marks)**

D-C

2 Which of these substances could **not** be used to neutralise an acid? Put a cross (☒) in the box next to your answer.

☐ **A** calcium oxide
☒ **B** sodium chloride
☐ **C** potassium hydroxide
☐ **D** magnesium carbonate

(1 mark)

B-A*

3 Calcium oxide combines with water to make calcium hydroxide $Ca(OH)_2$.

a) Write the balanced equation for this reaction.

$CaO + H_2O \rightarrow Ca(OH)_2$

(2 marks)

Guided

b) Calcium hydroxide is used by farmers to help neutralise acid soil. For example, it reacts with sulfuric acid (H_2SO_4) to form calcium sulfate ($CaSO_4$). Write the balanced equation for the reaction.

$Ca(OH)_2 + H_2SO_4 \rightarrow CaSO_4 + 2H_2O$ **(2 marks)**

c) Scientists measure the strength of acids using numbers on a scale. It is called the pH scale. A pH number higher than 7 is alkaline, lower than 7 is acidic and exactly 7 is neutral. Different plants prefer soils that have different pH numbers.

Range of pH numbers	Gardeners describe soil in this range as	Plants that like soil in this range
4.5–5.0	very strong acid	blueberries
5.1–5.5	strong acid	potato
5.6–6.0	moderate acid	carrot
6.1–6.5	slight acid	cucumber
6.6-7.0	neutral	lettuce
7.1–7.5	slightly alkaline	cabbage

i) Fill in the pH range for 'neutral' in the gardeners' table. **(1 mark)**

EXAM ALERT

ii) Describe how the scientists and gardeners disagree on the exact use of the word 'neutral'.

Gardeners say neutral is from 6.6-7.0, but scientsts says neutral is exactly 7.

(2 marks)

Students have struggled with exam questions similar to this – **be prepared!** ResultsPlus

Make sure you know what scientific words like 'neutralise' mean and can use them in the exam.

The importance of chlorine

E-C **1** The diagram shows the apparatus for producing chlorine using electricity.

hydrogen gas
chlorine gas
hydrochloric acid
– +
6V DC

a) State the name given to this process.

....

(1 mark)

b) State which type of current is supplied by the power source.

....

(1 mark)

Guided **c)** Write the word equation for the electrolysis of hydrochloric acid.

.... → hydrogen + ...

(1 mark)

d) Explain why this experiment is carried out in a fume cupboard.

....

....

(2 marks)

D-C **2** Which of these is the formula for chlorine gas? Put a cross (☒) in the box next to your answer.

☐ **A** Cl ☐ **B** Cl^2 ☐ **C** Cl_2 ☐ **D** Cg_2 **(1 mark)**

D-C **3** Name the raw material used to produce chlorine on a large scale.

....

(1 mark)

E-C **4** Describe a test to show that a gas is chlorine.

....

....

(2 marks)

B-A* **5** Chlorine is produced on a large scale. Explain one use for chlorine.

....

....

(2 marks)

B-A* **6** Write the balanced equation for the decomposition of dilute hydrochloric acid by electrolysis. Include state symbols in the equation.

State symbols tell you the physical state of the each substance in a reaction. Solid is 's', liquid is 'l', gas is 'g' and dissolved in water is 'aq'.

....

....

(3 marks)

The electrolysis of water

1 a) Write a word equation for the electrolysis of water.

....

(1 mark)

b) Complete the table to show the state of each substance at room temperature.

Substance name	Formula	State
water	H_2O	liquid (l)
hydrogen	H_2	
oxygen	O_2	

(3 marks)

D-C

2 A student places a glowing splint into a test tube of a gas to test for a gas. There are two possible results.

a) The splint re-lights.

One student correctly says the name of the gas in the tube is ....

(1 mark)

b) The splint does not re-light. One student says the gas in the test tube must be carbon dioxide. Can the student be certain? Explain your answer.

....

....

(2 marks)

EXAM ALERT

When testing for gases, it is important to give the test *and* the result.

Students have struggled with exam questions similar to this – **be prepared!** ResultsPlus

E-C

3 Describe the test to show that a gas is hydrogen.

....

....

(2 marks)

B-A*

4 During the electrolysis of water, the ratio of the volumes of hydrogen to oxygen is 2:1. When dilute hydrochloric acid is electrolysed, the ratio of hydrogen to chlorine is 1:1.

By considering the balanced equations for these reactions, suggest why these are the ratios.

....

....

....

....

(4 marks)

Ores

D-C

1 a) Iron can be extracted from iron oxide either by electrolysis or by heating with carbon. Explain why iron is normally extracted by heating with carbon, rather than using electrolysis.

....

....

....

....

....

....

(2 marks)

most reactive ↑ least reactive

metal	method of extraction
potassium	electrolysis of a molten compound
sodium	
calcium	
magnesium	
aluminium	
zinc	heat an ore with carbon
iron	
tin	
lead	
copper	
silver	found as the uncombined element
gold	
platinum	

b) Barium is more reactive than aluminium but less reactive than sodium.

Look at the information on the reactivity series and explain which method of extraction would be used to extract barium metal.

....

....

(2 marks)

D-B Guided

2 Iron is extracted from iron oxide, Fe_3O_4, by heating it with carbon.

Write a balanced equation for this reaction.

Fe_3O_4 + ... C → .... Fe + ... CO_2

(2 marks)

B-A*

3 Some metals used in mobile phones and computers are only found in small amounts in most countries. In one country they are much cheaper to extract than other countries. This country suddenly restricts the sale of the metals to other countries.

'Suggest' questions ask you to apply your knowledge to an unusual situation. You do not need to remember ideas about the unusual situation.

a) Suggest, giving a reason, the effect that this restriction is likely to have on the price of the metals in the other countries.

....

....

(2 marks)

b) Suggest why each of the other countries might now start to mine for the metals in its own country.

....

....

(2 marks)

Oxidation and reduction

1 Choose words from the box to complete the sentences. Each word may be used once, more than once or not at all.

corrode	oxidation	oxidised	oxygen
reduction	reduced	rust	

When a metal corrodes a metal oxide is formed. This is an example of

.... because the metal has gained ....

When oxygen is removed from a compound, the compound is ....

Nickel oxide is heated with carbon to produce nickel. The nickel oxide has been

.... and the carbon has been ....

Most metals .... but only iron and steel rust.

(6 marks)

D-B

2 Explain how corrosion is linked to position in the reactivity series.

....

....

(2 marks)

You do not need to learn the reactivity series but you do need to know how its position in the reactivity series determines how easily a metal corrodes and how a metal is extracted from its ore.

D-C Guided

3 a) In each of the equations below, underline the reactant that has been oxidised and put a ring around the reactant that has been reduced.

(1 mark)

(copper oxide) + <u>carbon</u> → copper + carbon dioxide

carbon + magnesium oxide → magnesium + carbon dioxide

b) In the same way as in **a)**, suggest the substance oxidised and the one reduced in this equation.

zinc + copper oxide → copper + zinc oxide

(1 mark)

B-A*

4 Bauxite is an ore of aluminium oxide, Al_2O_3. After it is melted, the aluminium oxide is decomposed using electrolysis to form molten aluminium and oxygen.

a) Write the balanced equation for this reaction. Give the state symbols.

....

....

(3 marks)

b) State which material is reduced in this reaction.

....

(1 mark)

Recycling metals

D-C

1 The chart shows the percentage of five metals obtained by recycling objects made from those metals.

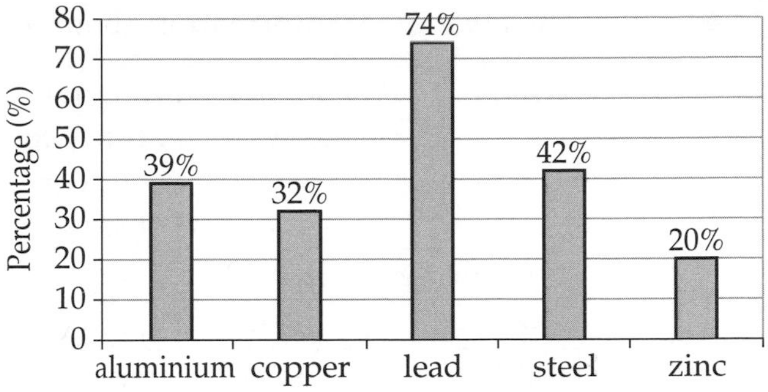

Metal	Percentage (%)
aluminium	39%
copper	32%
lead	74%
steel	42%
zinc	20%

Guided

a) The metals in the table are listed in alphabetical order.

Explain why analysing the data might be easier if the metals were drawn in the order of the reactivity series.

Then the order can be compared ....

This would tell us if the percentage recycled is ....

(2 marks)

b) The percentage of aluminium recycled is higher than the percentage of zinc recycled. Calculate the percentage difference between the two metals.

....

(1 mark)

C-B

2 The flow chart shows the stages in the purification of zinc from a zinc ore.

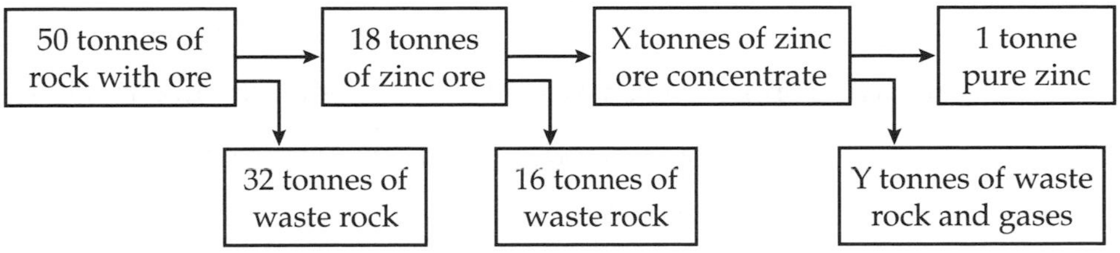

50 tonnes of rock with ore → 18 tonnes of zinc ore → X tonnes of zinc ore concentrate → 1 tonne pure zinc

50 tonnes of rock with ore ↓ 32 tonnes of waste rock

18 tonnes of zinc ore ↓ 16 tonnes of waste rock

X tonnes of zinc ore concentrate ↓ Y tonnes of waste rock and gases

a) Calculate the values of X and Y. Give the units.

X = ... unit .... Y = ... unit ....

(3 marks)

b) Calculate the percentage of zinc in the original 50 tonnes of rock containing ore.

Percentage of zinc in original rock.......................... %

(2 marks)

EXAM ALERT

c) Explain two reasons why it is better for the environment to recycle zinc than extract it from its ores. Use the chart to help you.

....

....

....

....

....

....

(4 marks)

Students have struggled with exam questions similar to this – **be prepared!** ResultsPlus

You need to give specific examples in an exam. Saying that recycling is 'better for the environment' is not detailed enough.

Alloys

D-C **1** The table gives information about four metals.

Metal	Density / kg/m^3	Relative ability to conduct electricity	Relative resistance to corrosion	Relative strength	Cost per tonne / £
aluminium	2700	good	good	high	1400
copper	8920	very good	good	high	5300
gold	19 300	very good	very good	low	34 000 000
steel	7800	poor	poor	very high	500

If you are given a table of data make sure that you use it in your answers.

Exam questions similar to this have proved especially tricky in the past – **be prepared!** ResultsPlus

You may be asked which properties of a metal are important for a particular use. Only give the relevant properties.

a) State two reasons why aluminium is sometimes used in the construction of bicycles, instead of steel. Use the information in the table.

....

....

(2 marks)

Guided **b)** Explain why copper is used instead of gold for electrical connecting wires.

Copper and gold are very good conductors of electricity and so both could be used for electrical wires but ....

(2 marks)

c) Explain why aluminium is used for electrical wires carried by pylons rather than copper.

....

....

(2 marks)

B-A* **2 a)** Explain what is meant by the term 'shape memory alloy'.

....

....

(2 marks)

b) State the name of the alloy made from nickel and titanium.

....

(1 mark)

c) Explain how this alloy is useful in the repair of collapsed blood vessels.

....

....

....

(3 marks)

Chemistry extended writing 2

Chlorine is manufactured in large factories. Every year, factories in the UK produce millions of tonnes of chlorine.

Describe how chlorine is produced in industry, and why it is important to make it safely and in such large quantities.

(6 marks)

You will be more successful in extended writing questions if you plan your answer before you start writing. This question asks you about how chlorine is made, what safety precautions need to be taken and why chlorine is so useful. Think about:

- the process of making chlorine on an industrial scale
- why safety is important and what precautions might be in place in the factory
- some examples of what chlorine is used to make.

Make sure that you use relevant technical terms when you write your answer.
In this answer, you should think about using the following terms: electrolysis, electrode, sodium chloride, seawater, toxic, bleach, PVC.

Chemistry extended writing 3

A man wants to buy some gold jewellery for his girlfriend. He sees two lovely rings in the jeweller's shop. One is made from 18 carat gold and the other is made from pure (24 carat) gold. Both rings weigh 50 grams.

Explain how the composition of the 18 carat ring is different to the 24 carat ring and why the two rings might have different properties.

(6 marks)

You will be more successful in extended writing questions if you plan your answer before you start writing.

This question is asking you to think about:

- how an alloy is different from a pure metal
- what 18 carat and 24 carat mean
- what else is in the rings, other than gold
- how the properties of the rings are affected by the different combinations of metals.

Remember that in an 'explain' question you are expected to say how and *why* for the statements that you make.

The question gives you the mass of the rings – 50 grams. You should make sure that you use this number in your answer. It will help you 'explain' your answer if you can use this information to say how the rings are different.

Your explanation of the different properties should also say what the different properties are *and why* they are different.

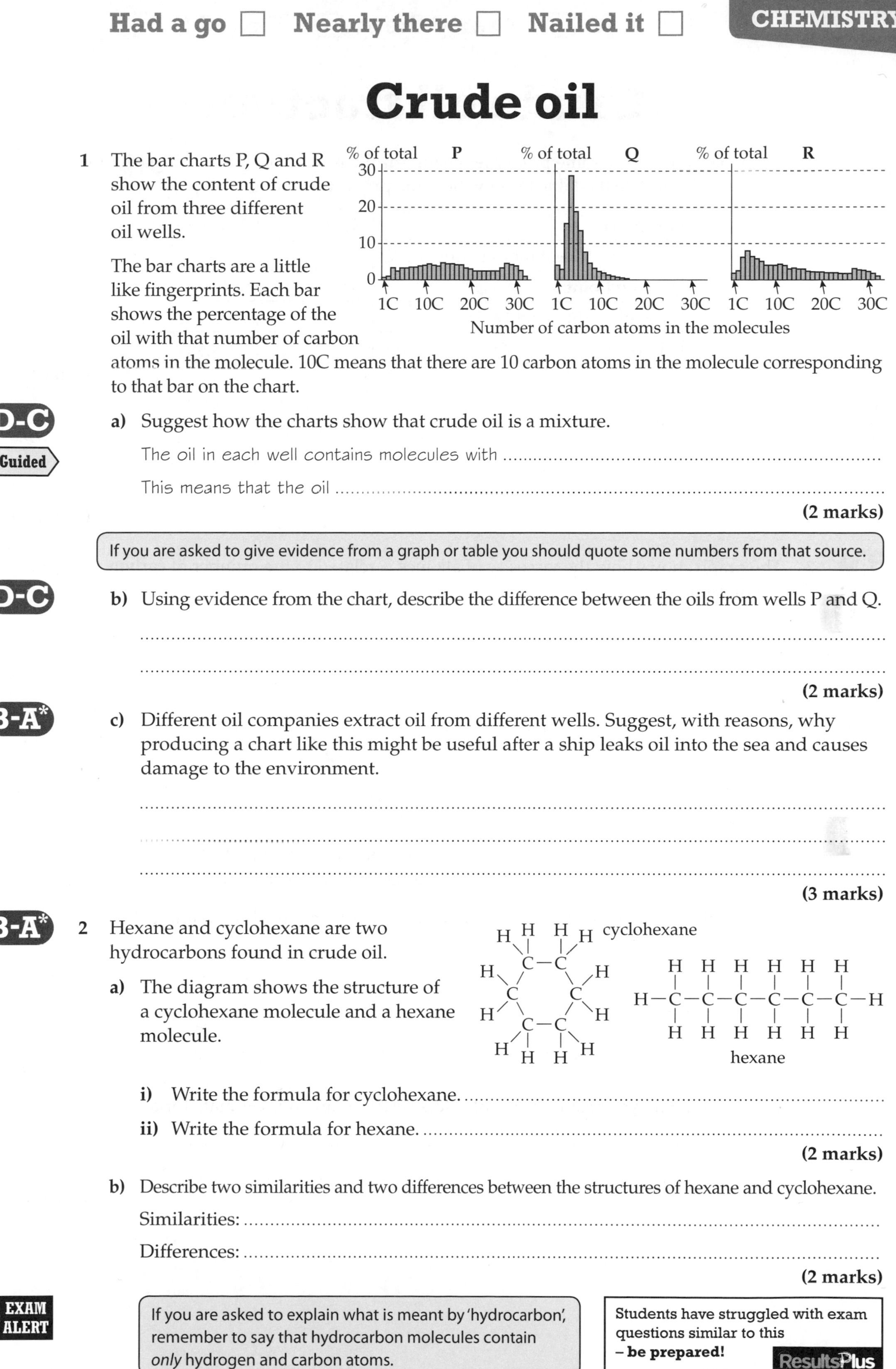

Crude oil

1 The bar charts P, Q and R show the content of crude oil from three different oil wells.

The bar charts are a little like fingerprints. Each bar shows the percentage of the oil with that number of carbon atoms in the molecule. 10C means that there are 10 carbon atoms in the molecule corresponding to that bar on the chart.

D-C **Guided**

a) Suggest how the charts show that crude oil is a mixture.

The oil in each well contains molecules with ....

This means that the oil ....

(2 marks)

If you are asked to give evidence from a graph or table you should quote some numbers from that source.

D-C

b) Using evidence from the chart, describe the difference between the oils from wells P and Q.

....

....

(2 marks)

B-A*

c) Different oil companies extract oil from different wells. Suggest, with reasons, why producing a chart like this might be useful after a ship leaks oil into the sea and causes damage to the environment.

....

....

....

(3 marks)

B-A*

2 Hexane and cyclohexane are two hydrocarbons found in crude oil.

a) The diagram shows the structure of a cyclohexane molecule and a hexane molecule.

i) Write the formula for cyclohexane. ....

ii) Write the formula for hexane. ....

(2 marks)

b) Describe two similarities and two differences between the structures of hexane and cyclohexane.

Similarities: ....

Differences: ....

(2 marks)

EXAM ALERT

If you are asked to explain what is meant by 'hydrocarbon', remember to say that hydrocarbon molecules contain *only* hydrogen and carbon atoms.

Students have struggled with exam questions similar to this – **be prepared!** ResultsPlus

Crude oil fractions

D-C Guided

1 The table shows how some properties change as the number of carbons in each molecule increases. Complete the table to show how the properties change.

Property	Gases	Bitumen
physical state at room temperature		liquid
boiling point	low	high
ease of setting alight		
viscosity		sticky

(4 marks)

D-B

2 Explain why oil companies spend lots of money on fractional distillation.

....

....

(2 marks)

3 The graph shows how the viscosity of alkanes is related to the number of carbon atoms in the molecules.

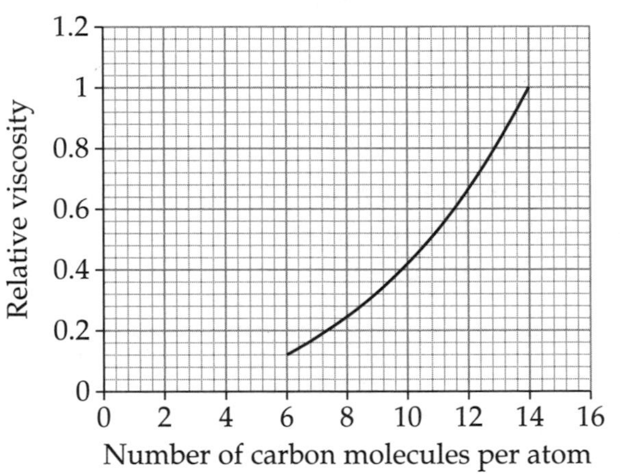

Draw lines on the graph if it helps you to work out the answer

a) Use the graph to estimate the relative viscosity of a molecule containing 9 carbon atoms.

....

(1 mark)

b) Estimate the relative viscosity of a molecule containing 5 carbon atoms.

....

(1 mark)

c) Explain why your estimate for **b)** is less certain than your estimate for **a)**.

....

....

(2 marks)

Combustion

D-C **1** Two students were investigating how much heat energy was produced when different hydrocarbons were burned. They used the heat from the hydrocarbons to heat water and then worked out how much energy was produced each time. They used hydrocarbons that had different numbers of carbon atoms in each molecule. Here are their results.

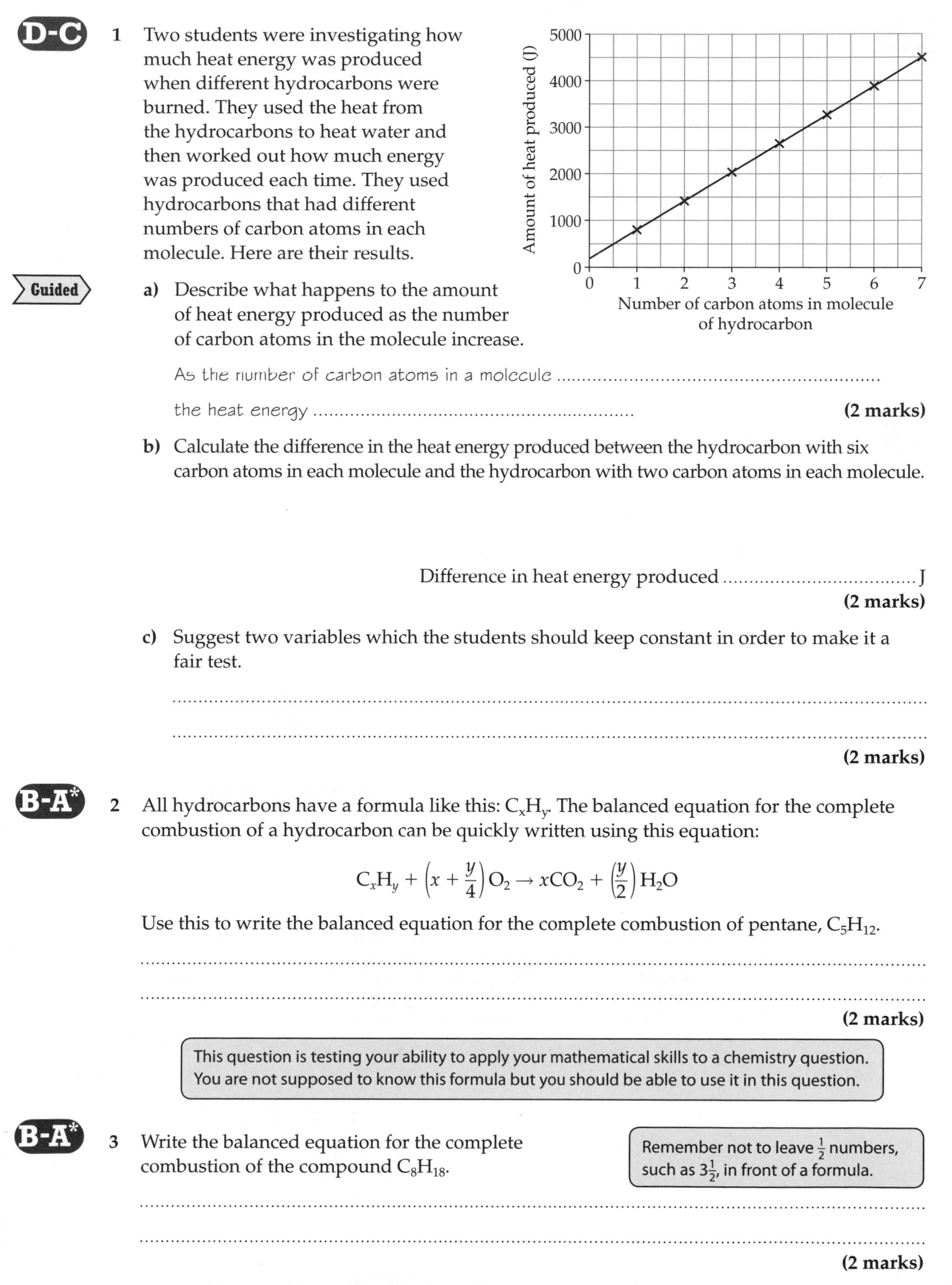

Guided **a)** Describe what happens to the amount of heat energy produced as the number of carbon atoms in the molecule increase.

As the number of carbon atoms in a molecule ....

the heat energy .... **(2 marks)**

b) Calculate the difference in the heat energy produced between the hydrocarbon with six carbon atoms in each molecule and the hydrocarbon with two carbon atoms in each molecule.

Difference in heat energy produced .... J

(2 marks)

c) Suggest two variables which the students should keep constant in order to make it a fair test.

....

....

(2 marks)

B-A* **2** All hydrocarbons have a formula like this: C_xH_y. The balanced equation for the complete combustion of a hydrocarbon can be quickly written using this equation:

$$C_xH_y + \left(x + \frac{y}{4}\right)O_2 \rightarrow xCO_2 + \left(\frac{y}{2}\right)H_2O$$

Use this to write the balanced equation for the complete combustion of pentane, C_5H_{12}.

....

....

(2 marks)

This question is testing your ability to apply your mathematical skills to a chemistry question. You are not supposed to know this formula but you should be able to use it in this question.

B-A* **3** Write the balanced equation for the complete combustion of the compound C_8H_{18}.

Remember not to leave $\frac{1}{2}$ numbers, such as $3\frac{1}{2}$, in front of a formula.

....

....

(2 marks)

Incomplete combustion

E-C 1 Explain why rooms that contain a fire need a constant supply of fresh air.

....

....

(2 marks)

B-A* 2 a) Explain the effect of carbon monoxide on the body.

....

....

(2 marks)

b) Write the word equation for the incomplete burning of methane, CH_4.

....

....

(2 marks)

B-A* 3 The diagram shows a bird's nest blocking the inlet and outlet pipes of the boiler in a house.

Explain how the nest affects the safety of the people who live in the house.

....

....

....

....

....

....

....

....

....

(3 marks)

B-A* 4 Complete this balanced equation for the incomplete combustion of propane.

Guided

> Incomplete combustion can give different quantities of each product depending on how much oxygen is available. In this question some of the balancing has been done for you to make it easier.

$\underline{2}\ C_3H_8 + \ldots\ldots\ O_2 \rightarrow \underline{2}\ C + \underline{2}\ CO + \underline{2}\ CO_2 + \ldots\ldots\ H_2O$

(1 mark)

Acid rain

D-B

1 Acid rain can make lakes and rivers become more acidic. This affects fish and the insects they feed on. Frogs are able to live even in the most acidic water.

The science subjects often work together.

Describe what the chart shows, using at least snails and trout as examples from the chart.

species found	neutral					more acidic
crayfish	✓	✓	✓	✗	✗	✗
frogs	✓	✓	✓	✓	✓	✓
mayfly	✓	✓	✓	✗	✗	✗
perch	✓	✓	✓	✓	✓	✗
snails	✓	✓	✗	✗	✗	✗
trout	✓	✓	✓	✓	✗	✗

....

....

.... **(2 marks)**

D-B

2 Acid rain affects the acidity of soil. Explain how farmers change the acidity of their soil by adding lime (calcium carbonate) to it.

....

....

(2 marks)

EXAM ALERT

Students have struggled with exam questions similar to this – **be prepared!** ResultsPlus

Make sure you read all the information in the question carefully. A recent exam question asked how passing waste gases from power stations through calcium carbonate helps to reduce the amount of acid rain. Many students said that the calcium carbonate was added to the clouds, even though the question told them how the calcium carbonate was used!

B-A*

3 The picture shows a gravestone made from marble. A neighbouring gravestone made from granite is easy to read. They were both put up at the same time.

IN
MEMORY OF
OUR BELOVED FATHER
WILLIAM TELFORD
DIED 16TH DECEMBER 1983
AGED 67 YEARS
REST IN PEACE

Explain the evidence supporting the view that the region suffers from acid rain.

....

....

....

(3 marks)

B-A*

4 Explain how a coal-fired power station can cause acid rain in a lake 50 miles away from it.

....

....

....

(3 marks)

Climate change

D-C **1** The graph shows how the concentration of carbon dioxide and the Earth's temperature have changed over the last 160 years.

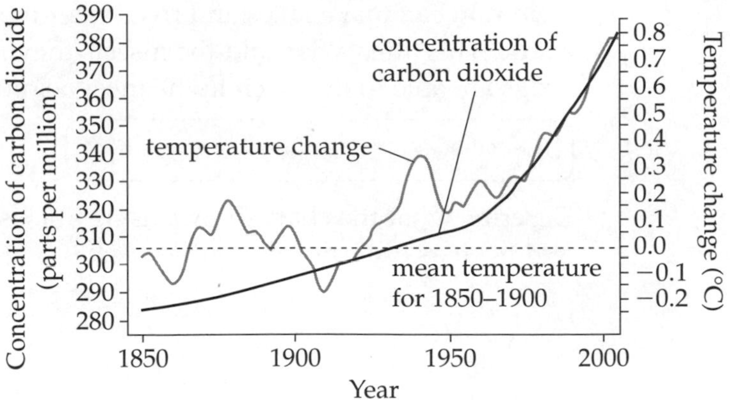

a) Describe how the Earth's mean temperature has changed in this time.

....

....

.... **(1 mark)**

b) Describe how the concentration of carbon dioxide has changed in this time.

....

(1 mark)

c) Estimate the year from which the temperature change has always been positive.

....

(1 mark)

d) State the possible relationship between temperature change and change in carbon dioxide levels.

....

....

(1 mark)

e) The value 0.8°C seems small. Suggest why it is important.

> If you are asked to evaluate something you need to use the evidence to make a judgement.

....

....

(2 marks)

D-C **2** Explain how burning fossil fuel in the UK makes the Earth warmer.

....

....

....

(3 marks)

B-A* **3** Some fuels are described as 'carbon neutral'. The fuels release the same amount of carbon dioxide into the atmosphere when they burn as they absorbed when they grew. So carbon neutral fuels should not affect the climate. Explain why ethanol produced from sugar cane is thought to be more carbon neutral than petrol made from crude oil.

....

....

....

(3 marks)

Biofuels

D-B 1 The table shows what is needed to grow different crops which are then all used to make biofuels.

Crop	Carbon dioxide emission	Input of water	Input of fertiliser	Input of pesticides	Input of energy	Amount of land needed
corn						
sugarcane						
soy beans						
canola						

Key: black – highest impact, dark grey – medium impact, lighter grey – medium/low impact, white – least impact.

Compare the use of sugarcane and soy beans as resources for biofuels.

....

....

....

(2 marks)

D-B 2 Some people want to use biofuels instead of fossil fuels. Explain how growing crops to produce biofuels will affect the production of human food.

....

....

(2 marks)

3 The table shows the emission of gases (apart from carbon dioxide) and small solid particles from cars using B100 and B20 biofuels. B100 biodiesel is 100% biodiesel. B20 biodiesel contains 20% biodiesel and 80% normal diesel.

The values given are the percentages of the amount that would be obtained from normal diesel. For example, B100 emits only 52% of the carbon monoxide that the same amount of normal diesel would emit.

	B100	B20	Normal diesel
Total unburned hydrocarbons	33%	80%	100%
Carbon monoxide	52%	88%	100%
Small solid particles	53%	88%	100%
Nitrogen dioxides (harmful)	110%	98–102%	100%

Using this information, describe how using B100 or B20 biodiesel instead of normal diesel will affect the quality of the air in busy areas.

Using these fuels will .... *because they both release*

.... *. However the new fuels do* ....

....

(4 marks)

Choosing fuels

E-C **1** Explain one disadvantage of using a gas rather than a liquid as the fuel in a car.

When stating advantages and disadvantages, you need to be clear what is being compared. Here, the question asks for one disadvantage of using a gas as compared to a liquid.

....

....

(2 marks)

B-A* **2** The table gives the energy released by 1 kg of different fuels.

Fuel	Energy released (MJ/kg)	Density at atmospheric pressure (kg/m^3)
diesel	44.8	850
hydrogen	141.8	0.089
methane	55.5	0.72
petrol	47.3	740

Using hydrogen in cars may help prevent climate change. If hydrogen is used in cars then less fossil fuels need to be burned.

Guided **a)** Write the balanced equation for the combustion of hydrogen. **(2 marks)**

The reactants and products have been given here but you need to balance it.

.... H_2 + ... O_2 → .... H_2O

b) State two advantages and two disadvantages of using hydrogen as a source of fuel for cars.

....

....

....

....

(4 marks)

c) Hydrogen seems to be quite a good source of energy. Suggest, with an economic reason, why hydrogen is not more commonly used.

....

....

....

(2 marks)

Alkanes and alkenes

1 Explain why alkane molecules are referred to as 'saturated' but alkene molecules are 'unsaturated'.

....

....

(2 marks)

2 Box P contains the formula for methane and a diagram to show how the atoms are arranged. Complete the names, formulae and the structures of the next three molecules in the alkane family.

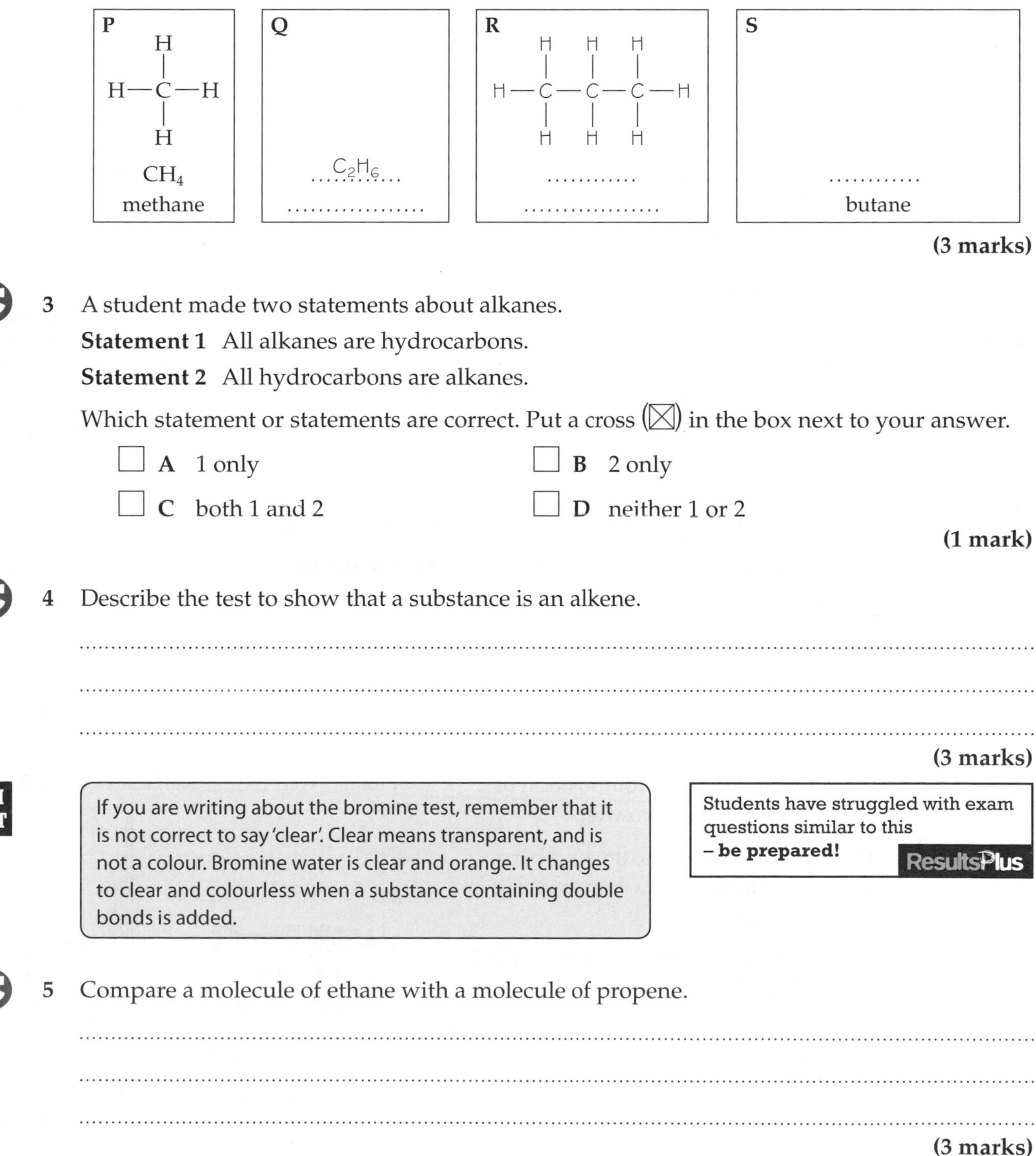

(3 marks)

3 A student made two statements about alkanes.

Statement 1 All alkanes are hydrocarbons.

Statement 2 All hydrocarbons are alkanes.

Which statement or statements are correct. Put a cross (☒) in the box next to your answer.

☐ **A** 1 only ☐ **B** 2 only

☐ **C** both 1 and 2 ☐ **D** neither 1 or 2

(1 mark)

4 Describe the test to show that a substance is an alkene.

....

....

....

(3 marks)

EXAM ALERT

If you are writing about the bromine test, remember that it is not correct to say 'clear'. Clear means transparent, and is not a colour. Bromine water is clear and orange. It changes to clear and colourless when a substance containing double bonds is added.

Students have struggled with exam questions similar to this – **be prepared!** ResultsPlus

5 Compare a molecule of ethane with a molecule of propene.

....

....

....

(3 marks)

Cracking

D-C 1 The diagram shows a cracking reaction.

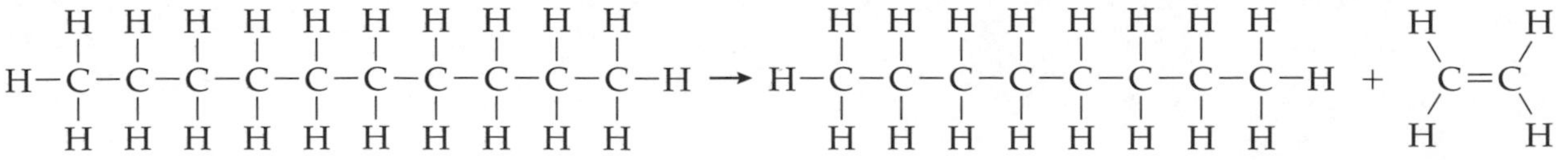

```
  H H H H H H H H H H             H H H H H H H H         H   H
  | | | | | | | | | |             | | | | | | | |          \ /
H-C-C-C-C-C-C-C-C-C-C-H  →  H-C-C-C-C-C-C-C-C-H  +   C=C
  | | | | | | | | | |             | | | | | | | |          / \
  H H H H H H H H H H             H H H H H H H H         H   H
```

longer alkane molecule   shorter alkane molecule   ....

a) Give the name of the second hydrocarbon formed in the reaction.

(1 mark)

b) State the balanced equation for this cracking.

....

....

(2 marks)

B-A* 2 By completing the balanced equation for this reaction, predict the formula and name for the missing product.

$C_{10}H_{22} \rightarrow C_7H_{16}$ + ...

The missing product is ....

(2 marks)

EXAM ALERT

Make sure you know the difference between fractional distillation and cracking. Fractional distillation is a physical separation process, and does not change any molecules. Cracking is a chemical reaction, and *breaks down* the molecules.

Students have struggled with exam questions similar to this – **be prepared!** ResultsPlus

B-A* 3 Crude oil is separated out into fractions by fractional distillation.

a) Explain why oil companies then take some of the fractions and put them through the process of cracking.

....

....

(2 marks)

b) The graphs show the composition of the crude oils from two wells.

On the horizontal axis, the numbers show the number of carbon atoms in the molecule.

Explain why oil from one of the wells is probably not sent for cracking.

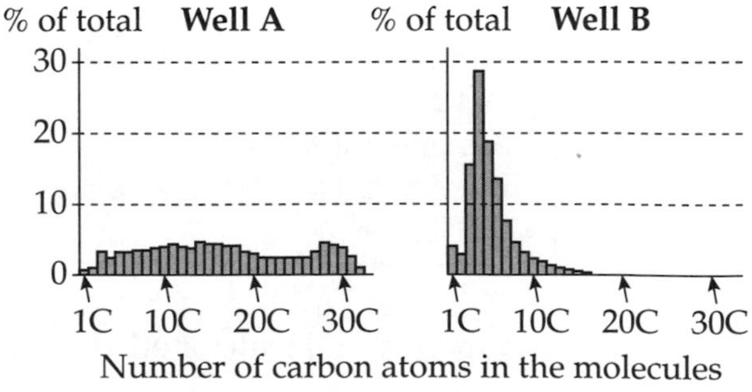

% of total **Well A**   % of total **Well B**

30, 20, 10, 0

1C 10C 20C 30C   1C 10C 20C 30C

Number of carbon atoms in the molecules

....

....

(2 marks)

Polymerisation

D-C Guided

1 a) Identify the monomers for the two polymers shown.

```
 F   F   F   F   F   F          H   H   H   H   H   H
 |   |   |   |   |   |          |   |   |   |   |   |
-C — C — C — C — C — C-        -C — C — C — C — C — C-
 |   |   |   |   |   |          |   |   |   |   |   |
 F   F   F   F   F   F          H   H   H   H   H   H
```

poly(...tetrafluoroethene...) poly(........................) **(2 marks)**

b) State two differences between a monomer and a polymer.

....

....

(2 marks)

D-C

2 The table shows different polymers and their uses.

Polymer	Properties
Poly(ethene)	flexible, cheap
Poly(propene)	flexible, shatterproof, has a high softening point
Poly(chloroethene) (PVC)	tough, cheap, longlasting, good electrical insulator
PTFE	tough, slippery

This question is asking you to *explain* so you need to say which polymer is used for windows but then give a reason *why* that polymer is used.

a) Explain which polymer is used to make window frames.

....

....

(2 marks)

b) Explain which polymer is used for coating skis.

....

....

(2 marks)

B-A*

3 a) The formula for chloroethene is C_2H_3Cl. Draw a molecule of chloroethene.

(2 marks)

b) Chloroethene is used to make a polymer. Draw a diagram to show how chloroethene forms the polymer. (Limit your drawing to three monomer molecules joining together.)

(2 marks)

Problems with polymers

1 Milk used to be delivered in glass bottles, which were collected and reused. Explain why it is better to reuse an object rather than recycle it.

....

....

....

(2 marks)

B-A*

2 The graph shows how the mass of plastic bottles recycled changed between 1994 and 2004.

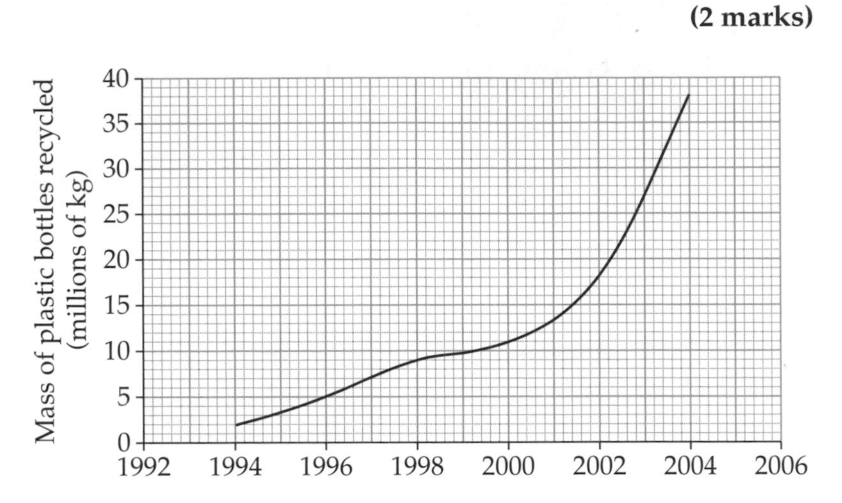

Guided

a) Calculate the rate of increase in the mass of bottles recycled between 1994 and 1996. State the units.

increase from 1994 to 1996 is .... millions of kg in two years

rate of increase = ... millions of kg per year **(3 marks)**

b) What was the average rate of increase in the mass of bottles recycled between 2002 and 2004?

Average rate of increase .... units .... **(3 marks)**

c) Suggest two possible reasons for this difference between the recycling in these periods (i.e. 1994 to 1996 and 2002 to 2004).

....

....

(2 marks)

d) Explain why it is not easy to predict, from this data, the mass of bottles recycled in 2020.

....

....

(2 marks)

B-A*

3 China imports more than 3000 million kg of plastic each year to recycle.

Discuss whether it is environmentally better for Britain to send plastic waste to China to be recycled rather than to put the waste in landfill sites in Britain.

....

....

....

(3 marks)

Chemistry extended writing 4

Venezuela is one of the largest oil-producing countries in the world. The oil that it produces is called 'heavy crude oil'. It gets the name 'heavy crude' because it contains a large proportion of the heavy fractions, such as bitumen. 'Heavy crude' contains very few of the 'top' fractions – those that come from the top end of the fractionating column during fractional distillation.

Explain how the process of cracking is used to make a Venezuelan 'heavy' crude oil more useful for customers.

(6 marks)

You will be more successful in extended writing questions if you plan your answer before you start writing.

Here the question asks you to explain what cracking is and why it is important. Think about:

- What is meant by the term 'cracking'?
- What does the cracking process do?
- How is cracking carried out?
- Which fractions of crude oil are customers most likely to want?
- Why does cracking help to make Venezuelan oil more useful?

Chemistry extended writing 5

The petrol we use in cars is a non-renewable fuel. This means that it will eventually run out. Scientists are trying to develop renewable fuels for cars. Two alternatives that they have investigated are ethanol (a biofuel) and hydrogen.

Compare how suitable these two fuels are as a replacement for petrol in cars.

(6 marks)

You will be more successful in extended writing questions if you plan your answer before you start writing.

This question is about fuels, and how replacement fuels – ethanol and hydrogen – are different from petrol. You need to think about:

- how easily the fuels burn and how much energy they release
- how they are made
- how easily they are stored and transported – this depends on their physical state
- whether they are more or less polluting than petrol.

It is tempting in a 'compare' question to write only about the things that are different between the things being compared – but you are still comparing, even if the properties are the same!

The best way to 'compare' is to think about the properties and qualities of each fuel and to write about how BOTH fuels behave.

Remember that you can write your answer in any way that you think best. One of the best ways to answer a 'compare' question is to use a table.

The Solar System

D-C **1** This diagram was drawn by Galileo when he looked at the planet Jupiter. Which of the following answers was the conclusion that Galileo came to? Put a cross in the box (☒) next to your answer.

East West January 7th, 1610	January 8th	[CLOUDY] January 9th
January 10th	January 11th	January 12th
January 13th	[CLOUDY] January 14th	January 15th

☐ **A** There were spots on his telescope.

☐ **B** The spots were moons in orbit around Jupiter.

☐ **C** The spots were moons in orbit around the Earth.

☐ **D** The spots were planets in orbit around the Sun.

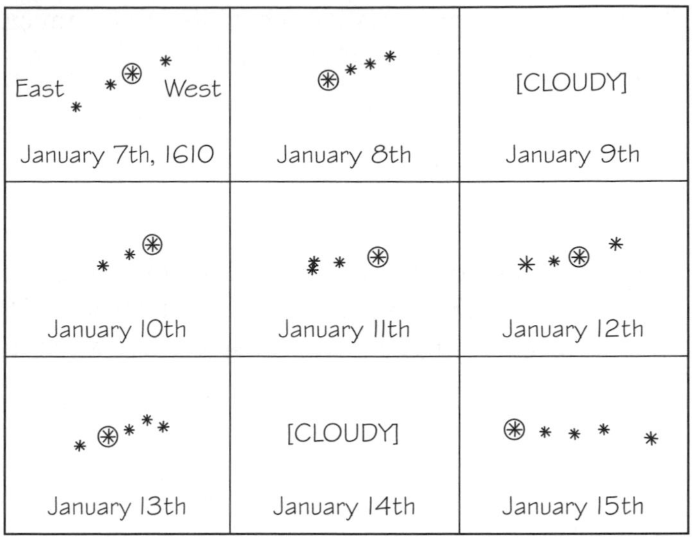

(1 mark)

D-C Guided **2** **a)** Explain why we can see more stars if we look through a telescope than if we use our naked eyes.

A telescope collects ....

With a telescope we can see ....

(2 marks)

b) Explain why astronomers today find it useful to photograph distant galaxies.

Think about the difficulties of recording observations of very distant objects.

....

....

....

(3 marks)

B-A* **3** **a)** Describe the differences between the geocentric model of the Universe and the heliocentric model.

Lots of different people have suggested models to explain how planets and stars are arranged in space. A geocentric model of the Universe was suggested by Ptolemy, and Copernicus was one of those people who suggested a heliocentric model.

....

....

(2 marks)

b) The way we look at space has changed since Ptolemy and Copernicus. State the type of wave that was used to make early observations and the types of waves scientists can use today.

....

....

(2 marks)

Reflection and refraction

1 Give two ways in which the *virtual* image in a magnifying glass is different from the *real* image formed by a lens.

A virtual image is .... but a real image is ....

A virtual image is .... but a real image is ....

(2 marks)

2 A student wants to find the focal length of a converging lens using just a sheet of paper and a ruler. She places the lens at the end of the ruler closest to a window. Then she holds the sheet of paper up beside the lens and moves it along the ruler.

a) State why the student pointed the lens at a window.

....

(1 mark)

b) State what the student was looking for when she moved the paper along the ruler.

....

(1 mark)

c) State how the student can measure the focal length.

....

(1 mark)

B-A*

3 A ray of light travelling through air hits a sheet of glass at an angle to the normal.

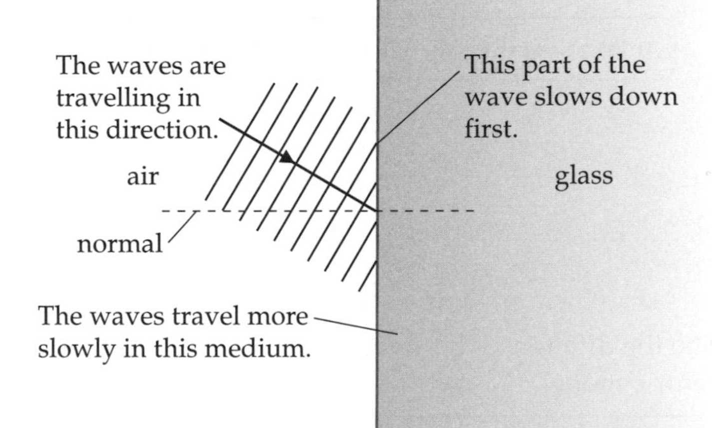

a) Complete the diagram to show what happens as the light waves enter the glass.

(3 marks)

The key points to show on the diagram are the change in direction towards the normal and the decrease in wavelength as the wave moves into the medium, where it travels more slowly.

Students have struggled with exam questions similar to this – **be prepared!** ResultsPlus

b) Explain why refraction occurs whenever a wave crosses a boundary between materials.

....

....

(2 marks)

Telescopes

E-C Guided

1 An astronomer looks through the eyepiece of a reflecting telescope and sees an image of a planet that cannot be seen with the naked eye. Describe how the reflecting telescope forms the image.

Light from the planet enters the telescope and is ....

The secondary mirror ....

The eyepiece produces a ....

(3 marks)

E-C

2

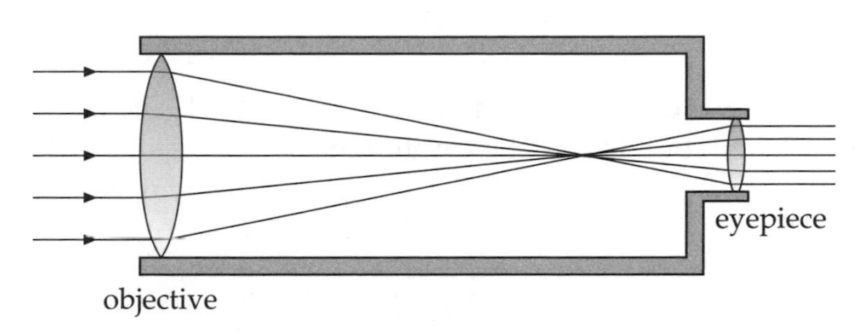

A simple refracting telescope, as shown in the diagram, is used to observe a star in the Milky Way.

a) Describe the image formed by the objective lens.

An image can be described by its size, whether it is the right way up or upside down and whether it is real or virtual.

....

....

....

(3 marks)

b) Explain why the image formed by the objective lens must be just closer to the eyepiece than the focus of the eyepiece lens.

....

....

(2 marks)

3 **a)** Describe one similarity and one difference between refracting and reflecting telescopes.

....

....

....

....

(4 marks)

b) All large telescopes, including the Hubble Space telescope, are reflecting telescopes. Suggest reasons why reflecting telescopes are used.

....

....

(2 marks)

Waves

1 A radio loudspeaker gives out sound waves. Describe what it is that travels from the radio to your ears.

Sound waves carry .... and .... from the loudspeaker to your ear.

(2 marks)

2 The diagram below shows a wave travelling through a medium.

a) Sketch a wave with the same wavelength but double the amplitude.

b) Sketch a wave with the same amplitude as in the diagram but with half the wavelength.

(2 marks)

c) The wave has a frequency of 50 Hz. Explain what this means.

....

....

(2 marks)

B-A* 3 When a wave travels through a material the average position of the particles of the material remains constant. Explain how this is correct for the type of waves found in:

a) a sound wave travelling through the air

....

....

(2 marks)

b) ripples travelling across the surface of a pond

....

....

Use relevant scientific terms wherever you can.

(2 marks)

Wave equations

You may find the following formulae useful:

wave speed = frequency × wavelength $v = f \times \lambda$

wave speed = $\frac{\text{distance}}{\text{time}}$ $v = \frac{x}{t}$

E-C Guided

1 A tap is dripping into a bath. Three drops fall each second, producing small waves that are 5 cm apart.

Calculate the speed of the small waves across the water. State the units.

The frequency of the waves (f) = ...

The wavelength of the waves (λ) = ...

Speed of waves .... unit ....

(3 marks)

The frequency is equal to the number of waves per second. The wavelength is equal to the distance they are apart. Use the first equation to work out the speed.

E-C Guided

2 Whales communicate over long distances by sending sound waves through seawater. It takes 20 seconds for the sounds to travel between two whales 30 kilometres apart.

Calculate the speed of sound through water in metres/second. (1 kilometre = 1000 metres)

EXAM ALERT

It is always a good idea to show your working. Even if you don't get the right answer you may gain some credit for having used the right method.

Students have struggled with exam questions similar to this – **be prepared!** ResultsPlus

Distance travelled by waves (in metres) = ... metres

Time taken = ...

Speed of sound ....m/s

(3 marks)

E-C

3 A girl sings with a frequency of about 200 Hz. The wavelength of the sound wave is 1.7 m. Calculate the speed of sound in air. State the units.

Speed of sound .... unit ....

(3 marks)

B-A*

4 A satellite sends signals to your TV using radio waves. It takes 0.12 s for the radio waves to travel from the satellite to your TV. The speed of light is 300 000 km/s.

Calculate the distance of the satellite above the Earth in kilometres.

....

....

....

(3 marks)

Beyond the visible

1 In 1801 Johann Ritter discovered what he called 'chemical rays' from the sun that were invisible and turned silver chloride black.

Describe an experiment to show that Ritter's 'chemical rays' are ultraviolet light.

Split up sunlight using a prism and ....

Measure the time taken for ....

Light beyond the violet will ....

(3 marks)

D-C 2 Below are some statements about work that William Herschel did in 1800.

A: Herschel thought that each colour of light produced a different amount of heat.

B: He used a prism to split sunlight into the colours of the spectrum and arranged for each colour to fall on a separate thermometer.

C: After repeated experiments he recorded that temperatures rise on average over 3°C at the red end and about 1°C at the blue end.

D: When the Sun moved in the sky a thermometer below the red end of the spectrum showed an even bigger temperature rise even though Herschel could not see any light falling on it.

E: Herschel checked the observation and decided that there must be invisible light beyond the red end of the visible spectrum.

F: He sent a record of his experiment to the Royal Society in London.

> Remember that observations are what you see and measure, predictions are what you think will happen before you do the experiment, and conclusions are what you think the observations mean.

Write down the letter of the sentence that describes:

a) a conclusion that Herschel drew from his experiment ....

b) the method that Herschel used in his experiment ....

c) the measurements that Herschel made ....

d) an hypothesis that Herschel was testing ....

e) how Herschel communicated his results ....

(5 marks)

3 The waves in the visible spectrum have wavelengths between 380×10^{-9} and 750×10^{-9} m, and frequencies between 400×10^{12} and 790×10^{12} Hz.

> It will be helpful to recall how wavelength and frequency differ for red and blue light

Suggest a wavelength and frequency for a wave of ultraviolet light.

....

....

(2 marks)

The electromagnetic spectrum

E-C

1 Microwaves and ultraviolet are types of electromagnetic radiation. Which of the following is a correct statement about microwaves and ultraviolet? Put a cross in the box (☒) next to your answer.

☐ **A** Microwaves have a higher frequency than ultraviolet.

☐ **B** Microwaves and ultraviolet are transverse waves.

☐ **C** Microwaves have a shorter wavelength than ultraviolet.

☐ **D** Microwaves and ultraviolet are longitudinal waves. **(1 mark)**

2 A candle gives out visible and infrared radiation. Gamma rays are given out by radioactive elements such as radium. Infrared, visible light and gamma rays are all part of the electromagnetic spectrum.

a) State two similarities between all the waves in the electromagnetic spectrum.

All parts of the electromagnetic spectrum are .... waves and they all .... **(2 marks)**

b) Explain why different parts of the electromagnetic spectrum have different properties.

....

(1 mark)

E-C

3 The chart represents the electromagnetic spectrum. Some types of electromagnetic radiation have been labelled.

longest wavelength/ lowest frequency

shortest wavelength/ highest frequency

← radio waves →← C →← infrared → B ←→← A →

ultra-violet rays (UV)

← gamma → rays

Name the three parts of the spectrum that have been replaced by letters in the diagram.

A: ....

B: ....

C: ....

> Remember the mnemonic – Red Monkeys In Vans Use X-ray Glasses. You may be given the spectrum the opposite way round in an exam.

(3 marks)

B-A*

4 The speed of electromagnetic waves in a vacuum is 300 000 km/s. A radio wave has a wavelength of 240 m. Calculate the frequency of the radio wave.

> You need to rearrange the formula provided $v = f \times \lambda$. Don't forget that the speed of the waves must be converted to m/s.

Frequency of the radio waves .... Hz

(3 marks)

Dangers and uses

E-C Guided

1 Read the postcard below from Sophie to her friend Kaylee. Sophie may have been in danger from ultraviolet, infrared and X-ray radiation. Identify the sources of radiation and describe the possible effects.

Hi Kaylee,
Had a great flight but it was really hot when we arrived – that's the tropics for you. Got on to the beach as soon as possible and lay in the Sun. The house we're staying in has got everything we need, including a toaster and electric cooker, so we're cooking for ourselves. I'm afraid I fell over a couple of days ago and hurt my wrist. I had to go to hospital to get an X-ray scan to see if I'd broken any bones. Heard all about the storms you had on the radio. Hope you're having fun.

Sophie x

St. Vincent $1
J. F. MITCHELL AIRPORT, BEQUIA

Kaylee Smith
12 Canal Lane
Birmingham
B2 3BH
England

She was exposed to ultraviolet radiation from the Sun, which could ....

....

She could have been exposed to infrared radiation from ....

....

She was exposed to X-rays from ....

....

(6 marks)

E-C

2 Below are the names of three types of electromagnetic radiation followed by four possible uses of each type. One of the uses given is **incorrect**. Put a cross in the box (☒) next to the incorrect use.

a) Infrared
☐ **A** night goggles ☐ **B** keeping warm
☐ **C** TV remote control ☐ **D** sun tan lamps

(1 mark)

b) Ultraviolet
☐ **A** disinfecting water ☐ **B** thermal imaging
☐ **C** security marking ☐ **D** fluorescent lamps

(1 mark)

c) Gamma rays
☐ **A** sterilising food ☐ **B** communicating with satellites
☐ **C** detecting cancer ☐ **D** treating cancer

(1 mark)

You need to remember the uses of these parts of the electromagnetic spectrum.

B-A*

3 Sportspeople are often given X-ray scans when they are injured, but X-rays are known to be harmful. Discuss the use of X-rays in the treatment of injuries.

....

....

....

(3 marks)

If you are asked to discuss something you need to include a conclusion in your answer.

Ionising radiation

E-C Guided

1 The figure is a hazard symbol showing that radioactive sources of ionising radiation may be present.

Suggest a reason why this symbol may be displayed in a hospital.

It is to warn people that ....

(1 mark)

C-A

2 In 1898 Marie and Pierre Curie tested samples of different substances with an instrument that detected if the air around the samples had been **ionised**. Look at the results they found.

Sample	Effect on detector
A	none
B	weak effect
C	strong effect
D	none
E	weak effect

a) Suggest a way that Marie and Pierre Curie could have made this experiment a fair test.

....

(1 mark)

b) List the substances they tested that were radioactive.

....

(1 mark)

Radioactive means that a substance gives out ionising radiation.

c) The results could be used to plan further experiments. Suggest two hypotheses that could be based on this data.

....

....

(2 marks)

First, decide whether the results show that radioactive samples all have the same effect on air, and then suggest *why* they may have different effects.

B-A*

3 Smoke alarms contain a source of alpha particles. Smoke alarms have a warning sign about the hazards of radiation. Ultraviolet lamps used by tanning salons also carry warning signs about the hazards of exposure to the rays. Compare alpha particles and ultraviolet rays.

....

....

....

....

(4 marks)

Physics extended writing 1

An astronomer views a star giving out blue light, using a reflecting telescope.

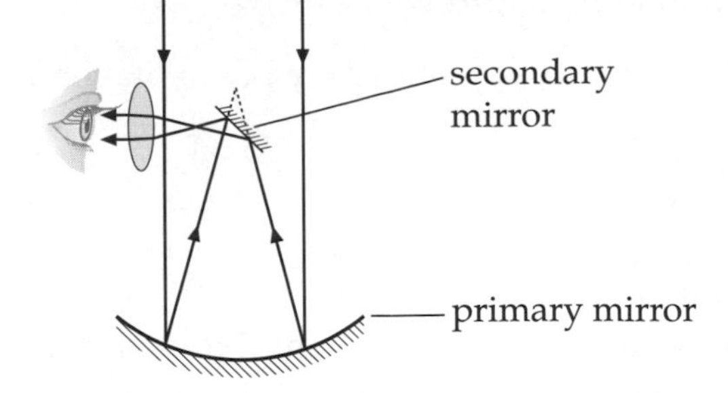

Explain what happens to the light from the star as it passes through the telescope and how this helps the astronomer see the star.

(6 marks)

You will be more successful in extended writing questions if you plan your answer before you start writing.

The question asks you to give a detailed explanation of what happens to a ray of light when it meets the different parts of the reflecting telescope. Think about:

- where the ray of light first enters the telescope and what happens to it
- what the image is like which is formed in the eyepiece
- how this helps the astronomer to see the star.

You should try to use the information given in the question.

It may help to draw a diagram to answer this question.

The Universe

D-C

1 The Cassini space probe was sent to the planet Saturn from Earth.

a) State which of the following is a true statement.
Put a cross in the box (☒) beside your answer.

☐ **A** Saturn orbits the Earth.

☐ **B** Saturn orbits the Sun.

☐ **C** Saturn orbits a moon.

☐ **D** The Sun orbits the Earth.

(1 mark)

b) Gliese 179 is a star in the Milky Way similar to the Sun. A planet has also been discovered orbiting Gliese 179. A space probe has not visited the planet orbiting Gliese 179.

i) Explain why space probes have visited Saturn but not the planets around Gliese 179.

....

....

(2 marks)

ii) Write the following in order of size starting with the smallest.

Saturn Gliese 179 the Moon the Universe the Milky Way

Remember that the Sun and Gliese 179 are both typical stars.

....

....

(2 marks)

C-B
Guided

2 Some people mix up the terms galaxy, Solar System and Universe. State what each term means.

The Solar System is ....

A galaxy, such as .... is a ....

The Universe is..

(3 marks)

B-A*

3 The Voyager 1 space probe was launched in 1977. It is now right at the edge of the Solar System. At the beginning of 2012 it took about 16 hours for radio signals to travel from the probe to Earth. Light from the Sun takes about 8 minutes to reach Earth. The Earth is 1 astronomical unit (AU) away from the Sun.

Estimate the distance of Voyager 1 from Earth in astronomical units.

An estimate is not a guess but uses approximate data to calculate an answer.

....

....

(2 marks)

Exploring the Universe

D-C Guided

1 A well-known nursery rhyme begins:
'Twinkle, twinkle little star, how I wonder what you are.'

a) Explain why stars seen in the night sky vary in brightness (twinkle).

Light from the stars passes through the air, which ....

....

(2 marks)

b) Describe the effect that light pollution has on the appearance of the night sky in a town.

The glow in the sky means that you see ....

(1 mark)

c) Suggest one solution to the problems referred to in parts **a)** and **b)**.

Move the telescope to ....

(1 mark)

EXAM ALERT

You need to be able to explain why some visible light telescopes are put into space.

Students have struggled with exam questions similar to this – **be prepared!** ResultsPlus

C-B

2

Radio telescopes detect radio waves

Optical telescopes detect visible light

The development of new astronomical instruments has led to many discoveries. State one advantage for each of the following developments.

a) Larger telescopes

....

(1 mark)

b) The use of cameras to take photographs

....

(1 mark)

c) Detectors for the whole of the electromagnetic spectrum

....

(1 mark)

B-A*

3 The Herschel Space Observatory (launched in 2009) orbits the Sun 1.5 million km from Earth. It carries instruments to detect and record infrared radiation with wavelengths between 55 and 625 micrometres (10^{-6} m). These instruments have detected water around a young star. Suggest why the water could not have been detected by instruments on Earth.

....

....

(2 marks)

Alien life?

D-C Guided

1 If you were looking at Earth from space, suggest the evidence that would reveal that there is life on Earth.

Earth is covered by oceans of ....

The Earth has plants that carry out .... and produce ....

Industry produces substances that ....

(3 marks)

EXAM ALERT

Students have struggled with exam questions similar to this – **be prepared!** ResultsPlus

Water only shows that life may exist on a planet or moon. It does not mean that life does exist.

D-B

2 The NASA Curiosity probe has been designed to land on Mars. Once on Mars, the probe moves around collecting evidence that may show that living organisms exist or did exist on the planet.

Suggest why a probe that lands on the planet may have a better chance of finding evidence of life than observations from space.

....

....

(2 marks)

D-B

3 SETI (the Search for Extraterrestrial Intelligence) is looking for evidence that other intelligent life-forms exist in the Universe.

a) Describe how SETI scientists are searching for evidence of intelligent life in the Universe.

....

....

(2 marks)

Remember that sound does not travel through space.

b) Some people say that as SETI has not found evidence of intelligent life elsewhere then it does not exist. Discuss whether this conclusion is justified.

....

....

(2 marks)

B-A*

4 In 1967 Jocelyn Bell discovered the first pulsar, a star that gives out regular bursts of radio waves. At first it was thought that the signals were evidence of extraterrestrial intelligence.

Explain why the discovery of other pulsars in other parts of the sky disproved this idea.

Think about the chances of different civilisations choosing the same method of communication.

....

....

(2 marks)

Life cycles of stars

E-C **1** Describe what causes a star at the red giant stage to change in size to form a white dwarf.

red giant

white dwarf

....

....

....

....

(2 marks)

D-C **2** The first stage in the life of a star is as a cloud of cold gas.

a) Name the force that pulls the gas into a compact ball.

....

(1 mark)

Guided **b)** Describe the energy changes that take place as particles of gas spiral inwards.

As particles start to spiral inwards the energy they have is ....

As the particles increase in speed they gain ....

When they collide this energy is transferred to ....

(3 marks)

C-B **3** A main sequence star like the Sun remains largely unchanged for up to 10 billion years.

Explain why the force that compressed the matter to form the star does not cause it to collapse during the main sequence.

> Remember that in a main sequence star the forces produced by two processes are balanced.

....

....

....

(3 marks)

B-A* **4** Compare the evolution of stars with mass equal to the Sun to the evolution of stars with considerably larger mass.

....

....

....

....

....

(5 marks)

Theories about the Universe

1 In the 1950s some scientists supported the Big Bang Theory and others the Steady State Theory.

a) Read the following statements and put a tick in the appropriate boxes to show which statements describe the two theories of the Universe.

Statement	Big Bang Theory	Steady State Theory
A The Universe started as a burst of energy from a tiny point.	✓	
B The Universe has always looked the same.		
C The Universe is expanding.		
D After the Universe formed particles came together to form galaxies.		
E New matter is constantly being formed.		✓

(2 marks)

b) State a reason why supporters of both theories accept the evidence that the light from distant galaxies is red-shifted.

....

(1 mark)

D-B 2 Explain why the Big Bang Theory eventually gained the support of most scientists.

A successful theory is one that explains all the available evidence.

....

....

(2 marks)

B-A* 3

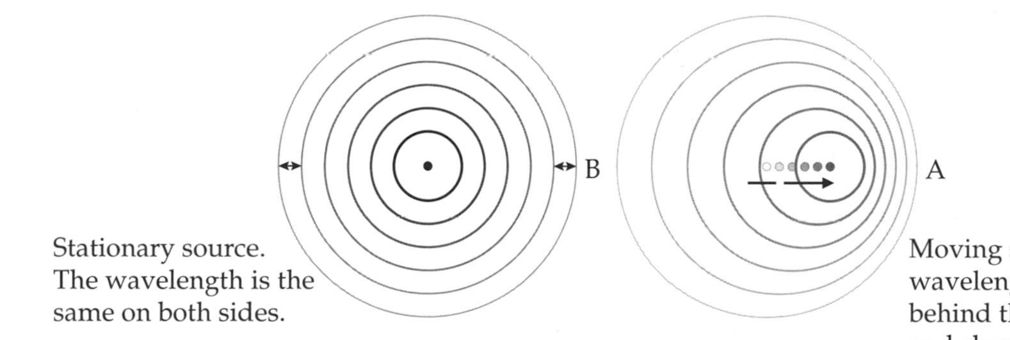

Explain why the sound of a moving siren is different in pitch than a stationary siren when you listen to it at points A and B.

A: ....

B: ....

(2 marks)

B-A* 4 Explain why Cosmic Microwave Background (CMB) radiation is evidence for the Big Bang Theory.

You need to think about what happened during and after the Big Bang.

....

....

(2 marks)

Infrasound

E-C

1 After the 2004 Indian Ocean earthquake it was reported that some animals moved inland before the tsunami hit the coast. Some scientists think this was because the animals reacted to infrasound waves produced by the tsunami.

Suggest evidence that would support this idea.

Read the paragraph carefully and decide which statements need evidence to support them.

....

....

(2 marks)

D-B

2 Volcanologists use infrasound produced during eruptions to determine the location of the volcano.

a) State what is meant by the term **infrasound**.

....

(1 mark)

b) Which of these sentences explains why infrasound is used this way?
Put a cross in the box (☒) next to your answer.

☐ **A** Infrasound can only be detected when it travels through the ground.

☐ **B** Infrasound does not travel through the ground.

☐ **C** Infrasound travels further through the ground than 'normal' sound.

☐ **D** Volcanic eruptions don't produce 'normal' sounds.

(1 mark)

Guided

c) Infrasound microphones can be set up to record the direction infrasound is coming from. Describe how this can be used to locate a volcanic eruption.

.... sets of microphones are set up ....

The sounds are recorded and lines ....

The eruption is at the point where ....

(3 marks)

B-A*

3 It is known that elephants communicate over distances of 10 km or more using an infrasound 'rumble' that travels through the air and the ground. Very few other animals can detect these sounds. Ordinary sounds only travel a few kilometres through the air. Evaluate the benefits to elephants of being able to communicate in this way

Think about reasons for communicating and the properties of sound waves.

....

....

(2 marks)

Ultrasound

D-C **1**

Frequency (Hz) → 20 human hearing range 20 000

A B C D

a) Choose the letter in the diagram that represents an ultrasound frequency.

....

(1 mark)

b) State **two** ways that animals make use of ultrasound.

....

....

(2 marks)

D-B Guided **2** Describe how ultrasound is used to build up a picture of a foetus in the womb.

The ultrasound probe is pressed against the woman's skin and ....

Some of the ultrasound passes through the foetus but some is ....

The machine measures ....

The computer uses the time taken ....

(4 marks)

B-A* **3** Ultrasound is used to produce an image of the structure of computer chips. The ultrasound waves are reflected by the different layers of material in the chip. An ultrasound signal is sent into the top surface of a chip and an echo is detected from a layer in the chip after 0.5 nanoseconds (0.5×10^{-9} s). The speed of sound in the computer chip is 8400 m/s. Calculate the distance of this layer below the surface of the chip.

Either use the powers of 10 in your calculation or watch the number of zeros carefully.

Distance .... m

(4 marks)

Seismic waves

D-B **1** After an earthquake, seismic waves are detected by seismometers. The graph shows how the difference in the arrival times of the two waves depends on the distance from the earthquake.

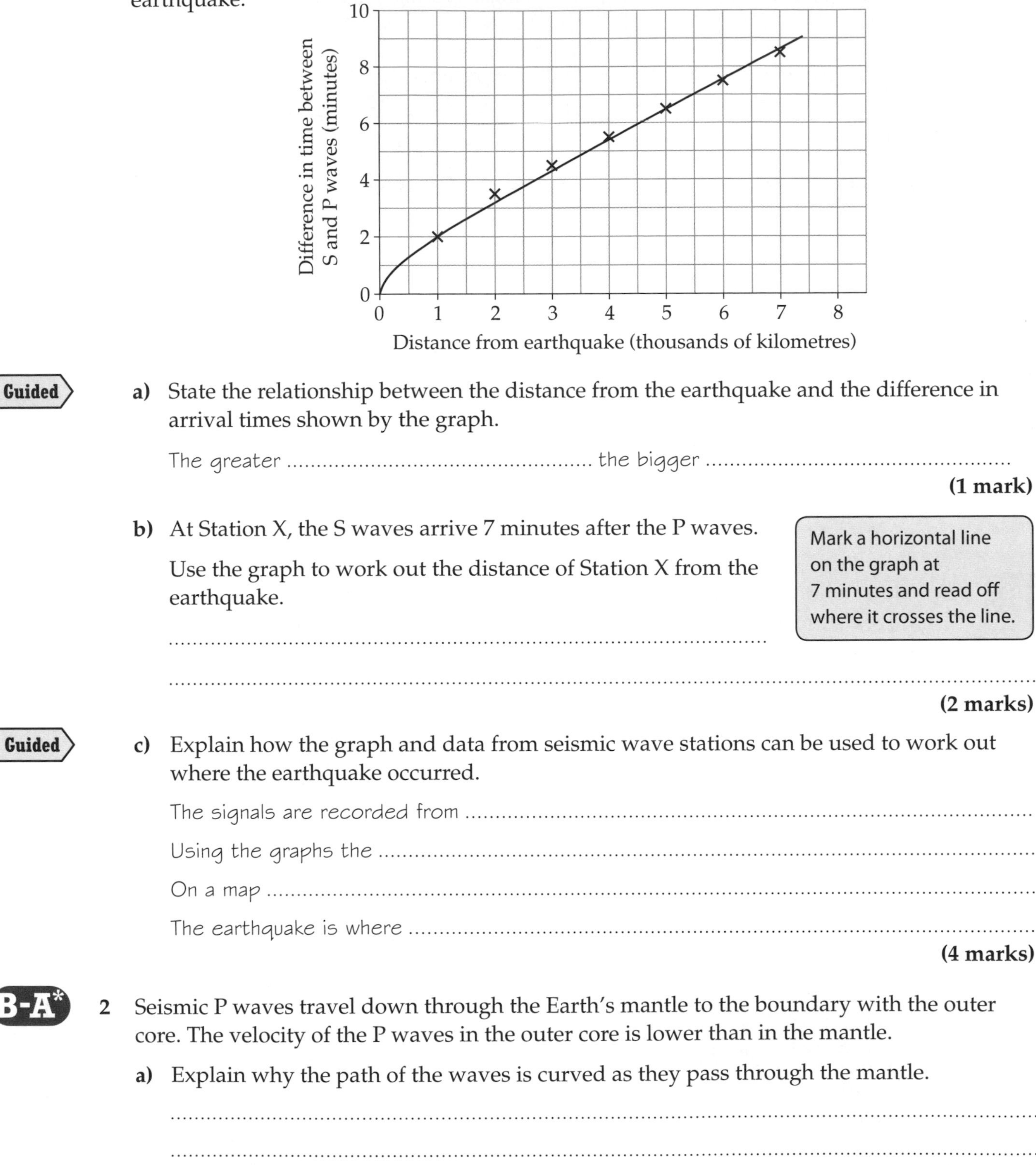

Guided

a) State the relationship between the distance from the earthquake and the difference in arrival times shown by the graph.

The greater .... the bigger ....

(1 mark)

b) At Station X, the S waves arrive 7 minutes after the P waves.

Use the graph to work out the distance of Station X from the earthquake.

Mark a horizontal line on the graph at 7 minutes and read off where it crosses the line.

....

....

(2 marks)

Guided

c) Explain how the graph and data from seismic wave stations can be used to work out where the earthquake occurred.

The signals are recorded from ....

Using the graphs the ....

On a map ....

The earthquake is where ....

(4 marks)

B-A* **2** Seismic P waves travel down through the Earth's mantle to the boundary with the outer core. The velocity of the P waves in the outer core is lower than in the mantle.

a) Explain why the path of the waves is curved as they pass through the mantle.

....

....

(2 marks)

b) Describe what happens to the P waves when they reach the boundary between the mantle and the outer core.

....

....

(2 marks)

Predicting earthquakes

D-C **1**

bungee cord
brick
sandpaper
plywood board

The diagram shows a **model**. The model can be used to demonstrate why scientists cannot predict when an earthquake may happen.

a) State what the brick and the elastic cord represent in the model.

....

....

(2 marks)

b) Describe what happens when the end of the elastic cord is pulled with more and more force.

....

....

(2 marks)

c) When students carry out this experiment they find that the force needed to make the brick start to move varies a lot. Suggest a reason why this is a good model of the way earthquakes happen.

....

(1 mark)

A model has to have some features or behaviour similar to the thing it is representing

D-B **2** The Earth's crust is broken up into various tectonic plates.

a) State the source of the force that makes tectonic plates move.

....

(1 marks)

b) Describe how tectonic plates are involved in causing earthquakes.

....

....

(2 marks)

B-A* Guided **3** A team of scientists announced that there was a 50% chance of a powerful earthquake in an area of 30 000 square kilometres of California during 2004. The authorities made no special plans in response to the prediction. There was no earthquake. Discuss whether the authorities' response was wise or lucky.

The authorities were wise because the prediction covered ....

Plans could ....

You may think that the authorities were lucky rather than wise so you could say so and explain why you think they were lucky.

(2 marks)

Physics extended writing 2

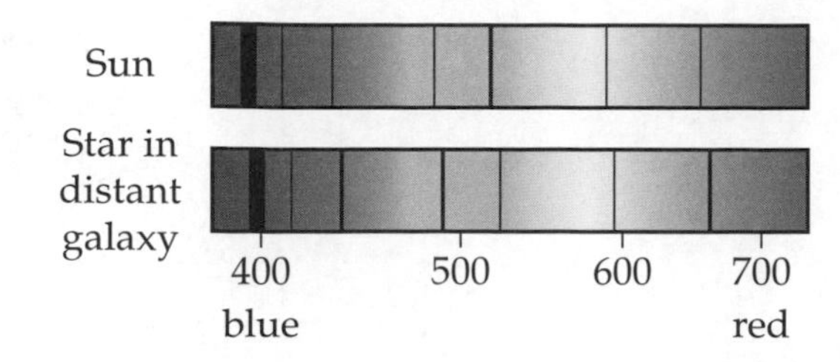

Spectra of light from the Sun and a star in a distant galaxy

The diagram shows the spectrum of light from the Sun and from a star in a distant galaxy. Data like this has been used to provide evidence for both the Big Bang Theory and the Steady State Theory of the Universe.

Explain how the differences between the two spectra provide evidence to support both the Big Bang and Steady State theories.

(6 marks)

You will be more successful in extended writing questions if you plan your answer before you start writing.

The question asks you to give a detailed explanation of the differences between the two spectra and how this provides evidence for both the theories. Think about:

- the differences between the spectra
- what causes the differences
- why we see red shift
- how red shift relates to each of the two theories about the origin of the Universe.

Think about what has happened to the light travelling from the distant star and why it is different.

Try to write your answers as full sentences using correct grammar, punctuation and spelling and using scientific words where appropriate.

Physics extended writing 3

An earthquake produces seismic waves that travel through the Earth. Scientists use seismometers placed at various points on the Earth's surface to detect and record the waves. The data allows scientists to work out where the earthquake occurred but also gives them information about the structure of the Earth.

Describe how the passage of seismic waves through the Earth gives information about the structure of the Earth.

(6 marks)

You will be more successful in extended writing questions if you plan your answer before you start writing.

The question asks you to describe how seismic waves travel through the Earth and how this can be used by scientists to tell them about the structure of the Earth. Think about:

- the types of seismic waves and any differences between them
- how seismic waves can give scientist information about the structure of the Earth
- what the structure of the Earth is
- how the structure of the Earth affects the seismic waves as they travel through the Earth.

It may help to draw a diagram to answer this question.

Renewable resources

1

1000 WIND TURBINES FOR LOCAL HILL TOPS

WINDTECH LTD wants to build up to one thousand wind turbines on the hills around Pentridwr. The local council is considering the proposal. Protest groups oppose the project, saying it will stop tourists visiting the region.

a) The wind turbine will provide an electric current. Describe the term **current.**

Current is ....

....

(2 marks)

b) Suggest two reasons that Windtech could give for going ahead with the project.

The advantages of using wind power are ....

....

(2 marks)

c) Suggest two reasons the protest groups could give for stopping the project.

The disadvantages of wind power are ....

....

(2 marks)

D-B 2 Some of the sources of renewable energy listed below are only available at certain times, while other sources can be used at any time.

hydroelectric tidal solar wind geothermal

a) Name the sources of renewable energy in the list that are always available.

....

....

(2 marks)

Think about how the weather affects some renewable energy sources.

b) Explain why it is an advantage to have a source of energy available at any time.

....

....

(2 marks)

B-A* 3 Renewable energy sources such as wind, solar and wave power use the 'free' energy from the Sun. Suggest why these renewable energy sources do not provide most of our electricity at present.

....

....

(2 marks)

Non-renewable resources

1

GOVERNMENT PLANS NEW NUCLEAR POWER STATIONS

The government has announced plans for a new series of nuclear power stations costing at least £3 billion each. The government says the new power stations are needed but opposition groups say that there are many reasons for not going ahead with the plans.

a) State two advantages of nuclear power

Nuclear power stations do not ....

Nuclear power stations can be switched on ....

(2 marks)

b) State two disadvantages of nuclear power

Nuclear power stations produce ....

Decommissioning a nuclear power station ....

(2 marks)

D-B

2 The following statements are about using fossil fuels as a source of energy. Explain which statements are advantages and which are disadvantages.

The question says 'explain', so as well as stating whether each statement is an advantage or disadvantage, you must also give reasons for your answer.

a) Fossil fuels are non-renewable.

This is a .... because ....

....

(2 marks)

b) Burning fossil fuels produces sulfur dioxide and nitrogen oxides.

This is a .... because ....

....

EXAM ALERT

Give details if you are answering a question about fossil fuels. Don't just say 'causes pollution', but link the pollutants to the effects they have.

Students have struggled with exam questions similar to this – **be prepared!** ResultsPlus

(2 marks)

c) Fossil fuel power stations can work at any time, day or night.

This is a .... because ....

....

(2 marks)

B-A*

3 Some people say that we have passed the time of 'peak oil'. After this time, the amount of crude oil extracted will slowly decrease and prices for fuel and electricity will rise rapidly. Other people say that we will not pass this peak until 2020.

Suggest why there is uncertainty as to when peak oil will be passed.

....

....

(2 marks)

Generating electricity

D-C

1 Look at the diagram.

movement of wire

induced current

ammeter

These answers can be quite short but try to use the technical terms for parts of the generator and principles of induction.

a) State what must be done to induce a current in the wire connected to the ammeter.

....

....

(1 mark)

b) State how a current in the opposite direction can be induced.

....

(1 mark)

c) State **three** things that can be done to increase the size of the current.

....

....

....

(3 marks)

D-B Guided

2 Power stations use very large generators to generate electricity. Compare the magnets used in the power station generators with a model generator (such as the one shown in question **1**) and give a reason for the difference.

The magnets used in power station generators are ....

because they ....

(2 marks)

B-A*

3 An electric generator produces an alternating current.

a) Sketch a line using the axes below, showing how the current changes as the coil of a generator is turned.

Current

Time

(2 marks)

b) Explain why this is described as an alternating current.

Think of how an alternating current and a direct current are different.

....

....

(2 marks)

Transmitting electricity

E-C

1 The National Grid transmits electricity from power stations to where it is needed at 400 000 volts (400 kV).

a) Explain why this voltage is used to carry electricity long distances.

....

....

(2 marks)

EXAM ALERT

Remember that increasing the voltage *decreases* the current.

Students have struggled with exam questions similar to this – **be prepared!** ResultsPlus

b) State one hazard of transmitting electricity at 400 000 V.

....

(1 mark)

D-B

2 Transformers are used at various places in the National Grid.

a) Describe the job that transformers do in the National Grid.

....

....

(2 marks)

b) Describe what happens to the electricity produced at 25 kV by a power station when it is transmitted to a house at 230 V many miles away.

....

....

....

(3 marks)

B-A*

Guided

3 A laptop computer needs a voltage of 19 V. It is connected to the 230 V mains electricity supply using a transformer with 380 turns on the secondary coil. Calculate the number of turns on the primary coil of the transformer.

You need to remember the equation for the voltage and turns in a transformer. Start by identifying which are the known and unknown quantities in the equation.

$\frac{V_p}{V_s} = \frac{n_p}{n_s}$ V_p = ..., V_s = ..., n_s = ...

Rearrange the equation and substitute the known values.

$n_p = \frac{V_p}{V_s} \times n_s =$ = ... turns

(3 marks)

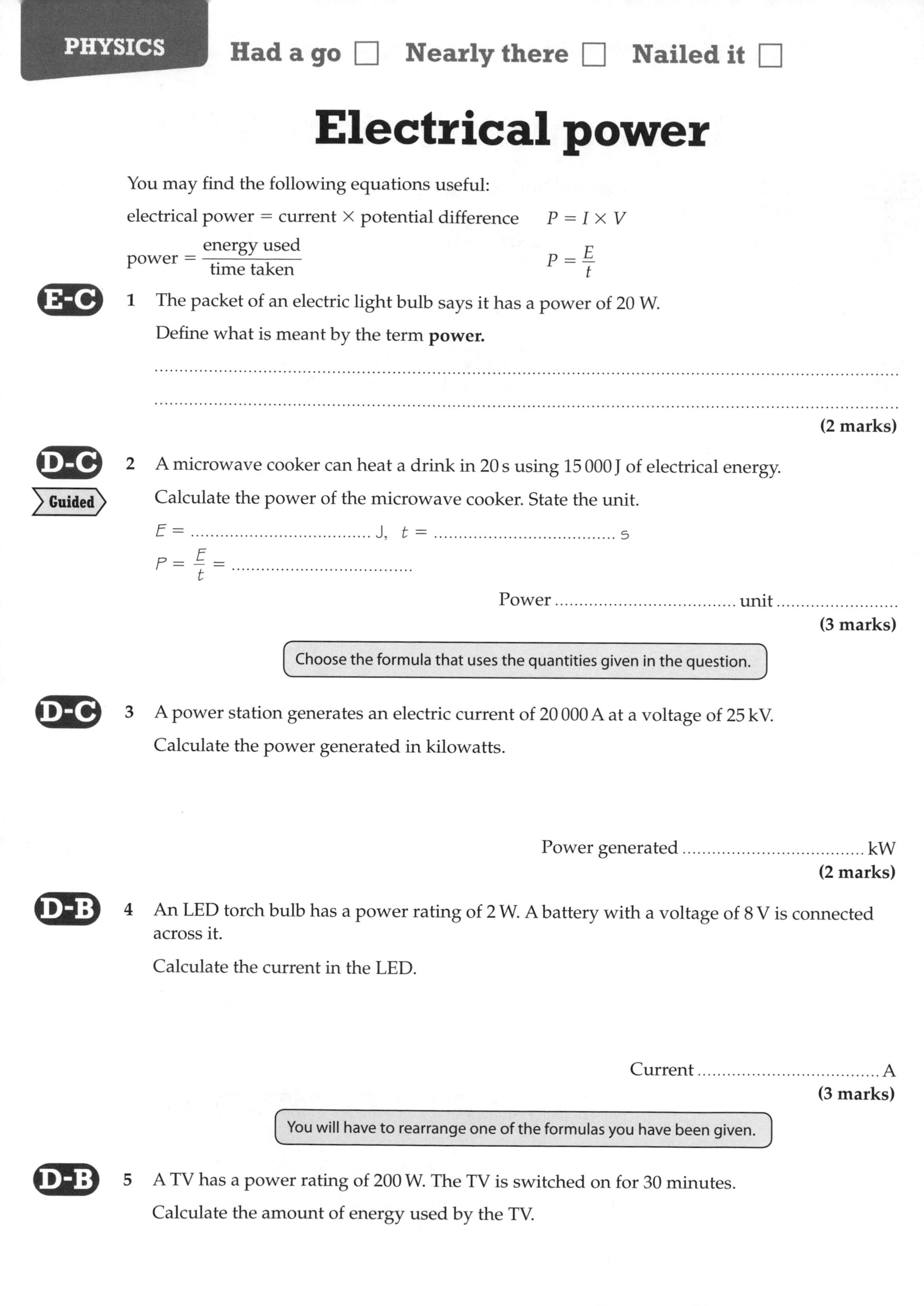

Electrical power

You may find the following equations useful:

electrical power = current × potential difference $P = I \times V$

$\text{power} = \frac{\text{energy used}}{\text{time taken}}$ $P = \frac{E}{t}$

E-C **1** The packet of an electric light bulb says it has a power of 20 W.

Define what is meant by the term **power.**

....

....

(2 marks)

D-C Guided **2** A microwave cooker can heat a drink in 20 s using 15 000 J of electrical energy.

Calculate the power of the microwave cooker. State the unit.

$E =$ .... J, $t =$ .... s

$P = \frac{E}{t} =$ ....

Power .... unit ....

(3 marks)

Choose the formula that uses the quantities given in the question.

D-C **3** A power station generates an electric current of 20 000 A at a voltage of 25 kV.

Calculate the power generated in kilowatts.

Power generated .... kW

(2 marks)

D-B **4** An LED torch bulb has a power rating of 2 W. A battery with a voltage of 8 V is connected across it.

Calculate the current in the LED.

Current .... A

(3 marks)

You will have to rearrange one of the formulas you have been given.

D-B **5** A TV has a power rating of 200 W. The TV is switched on for 30 minutes.

Calculate the amount of energy used by the TV.

Energy used .... J

(3 marks)

Paying for electricity

You may find this formula useful:

cost of electricity = power × time × cost of 1 kilowatt-hour

D-C

1 A student takes a shower for 12 minutes (0.2 hours). The shower is powered by an electrical heater with a power rating of 2 kilowatts. Her electricity supplier charges 10p for each kilowatt-hour used.

a) Put a cross in the box (☒) against the cost of her shower.

You need to do the calculation in rough.

☐ A £4 ☐ B 40p

☐ C 4p ☐ D 240p

(1 mark)

Guided

b) The student irons a blouse. The iron has a power rating of 1800 W. It takes her 6 minutes (0.1 hours) to iron the blouse. Electricity costs 10p for 1 kilowatt-hour.

Calculate the cost of ironing the blouse.

Power in kW = ...

Cost = power (in kW) × time (in hours) × cost (in pence)

Cost = ...

(3 marks)

2 A student uses the following appliances during one day.

- a 0.1 kW light for eight hours
- a 1 kW heater for six hours
- a 0.2 kW television for four hours

Power must be in kilowatts.

The cost of his electricity is 15p per kilowatt/hour. Calculate the cost of his electricity for the day.

Cost of electricity .... kW

(3 marks)

D-B

3 An electric cooker takes 15 minutes (0.25 hours) to heat up a pie. The cooker has a power rating of 3 kW. Electricity is charged at 12p per kilowatt hour. Calculate the cost of the electricity used.

Cost of electricity .... p

(2 marks)

B-A*

4 A student has one £1 coin left for the electricity meter in his flat. Electricity is charged at 10p per kilowatt-hour. Assuming he is using no other electrical appliances calculate how long the student can run his 2 kW heater for before he runs out of electricity.

Time .... hours

(3 marks)

Reducing energy use

D-C 1 A couple have moved into a two bedroom house which does not have cavity wall insulation. It is suggested that cavity wall insulation could reduce their energy costs by £100 a year. They decide to have it fitted at a cost of £250.

a) Explain how cavity wall insulation reduces energy costs.

....

....

(2 marks)

b) Calculate the payback time for installing the cavity wall insulation.

> Work out how many years of savings are needed to pay the initial cost.

....

....

(2 marks)

C-B **Guided** 2 A student's parents are trying to decide which model of freezer to buy to replace their old one that has broken. He draws up the table shown below to compare two models of freezer that are the same size.

Model	Cost to buy (£)	Electricity used in 1 year (kW-hr)
Ecool 1500	350	150
Freezemate 1.5	300	400

His parents are paying 12p per kilowatt-hour for their electricity.

Using the information above, explain which freezer is the most cost-effective over two years.

The costs of buying and running the freezers for one and two years are shown below.

Model	Cost to buy (£)	Cost of electricity each year	Total cost in 1st year	Total cost in 2 years
Ecool	350	150 × 12 = p		
Freezermate	300			

Therefore the best freezer to buy is ....

(4 marks)

B-A* 3 A house owner decides to fits solar panels to the roof of his house at a cost of £7200. The solar panels should produce 1200 kilowatt-hours of electricity during the year. The government pays 40p for each kilowatt-hour generated guaranteed for 25 years.

Discuss the house owner's decision.

....

....

....

(3 marks)

Energy transfers

D-C **Guided**

1 **a)** A marathon runner has a breakfast of cereal and toast before setting out on a training run. Draw an energy transfer chain for the energy used by the runner starting with breakfast.

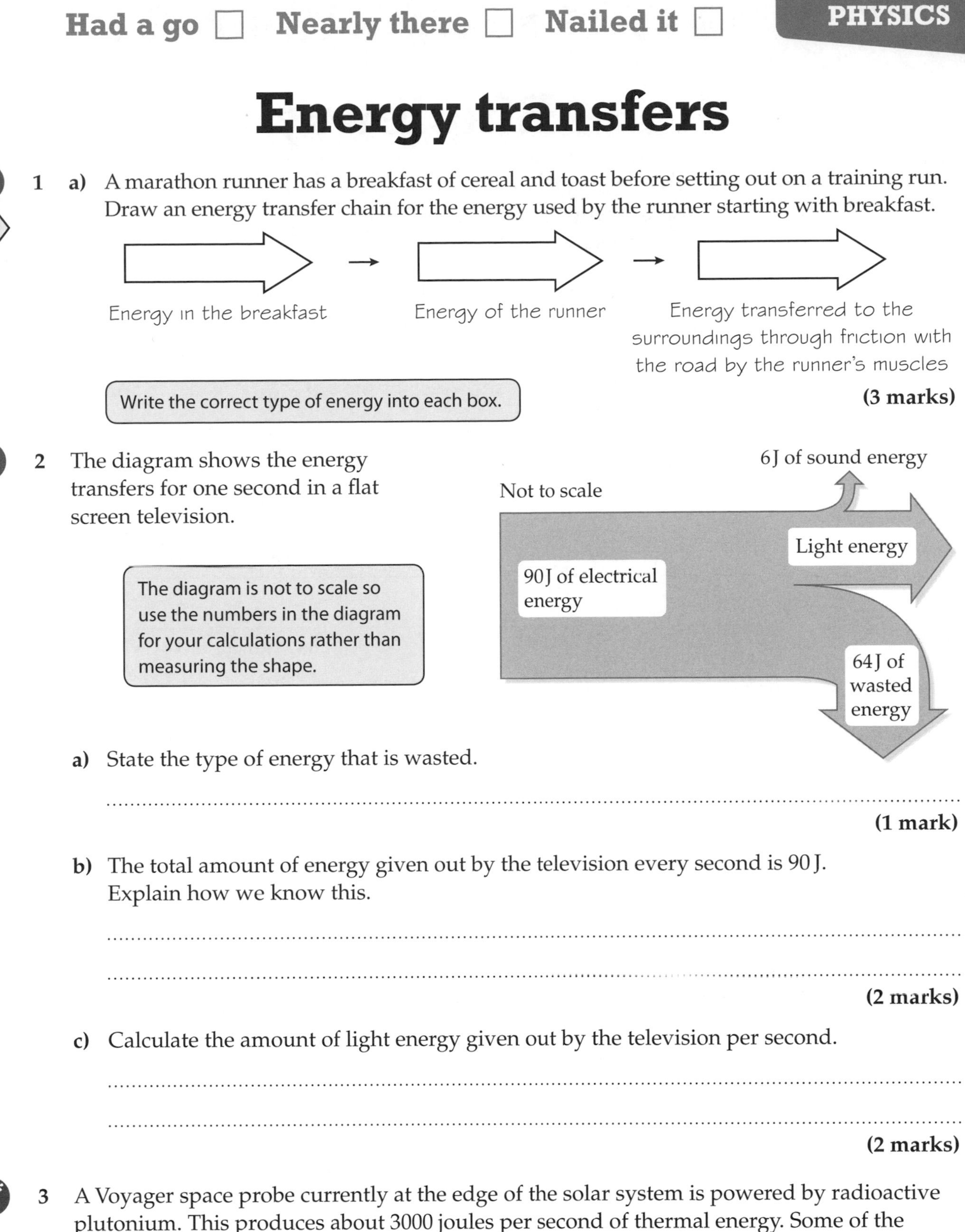

Write the correct type of energy into each box.

(3 marks)

D-B

2 The diagram shows the energy transfers for one second in a flat screen television.

The diagram is not to scale so use the numbers in the diagram for your calculations rather than measuring the shape.

a) State the type of energy that is wasted.

....

(1 mark)

b) The total amount of energy given out by the television every second is 90 J. Explain how we know this.

....

....

(2 marks)

c) Calculate the amount of light energy given out by the television per second.

....

....

(2 marks)

B-A*

3 A Voyager space probe currently at the edge of the solar system is powered by radioactive plutonium. This produces about 3000 joules per second of thermal energy. Some of the thermal energy is used to produce 300 joules per second of electrical energy. Most of the electricity is used by systems in the probe and is released as heat, but the radio transmitter sends signals to Earth using about 20 joules per second. Sketch an energy transfer diagram for the Voyager space probe.

(3 marks)

Efficiency

You may find the following formula useful

$$\text{Efficiency} = \frac{\text{(useful energy transferred by the device)}}{\text{(total energy supplied to the device)}} \times 100\%$$

D-C Guided

1 A crane lifts a box to the top of a building. The box has 1 million joules of gravitational potential energy when it is at the top of the building. The crane uses up fuel supplying 4 million joules. Calculate the efficiency of the crane.

The useful energy transferred = energy given to the box = ...

The energy supplied = the energy used in the fuel = ...

$$\text{Efficiency} = \frac{\text{(useful energy transferred by the device)}}{\text{(total energy supplied to the device)}} \times 100\% = \ldots\ldots\ldots\ldots$$

(2 marks)

Efficiency does not have a unit but is usually given as a percentage

2

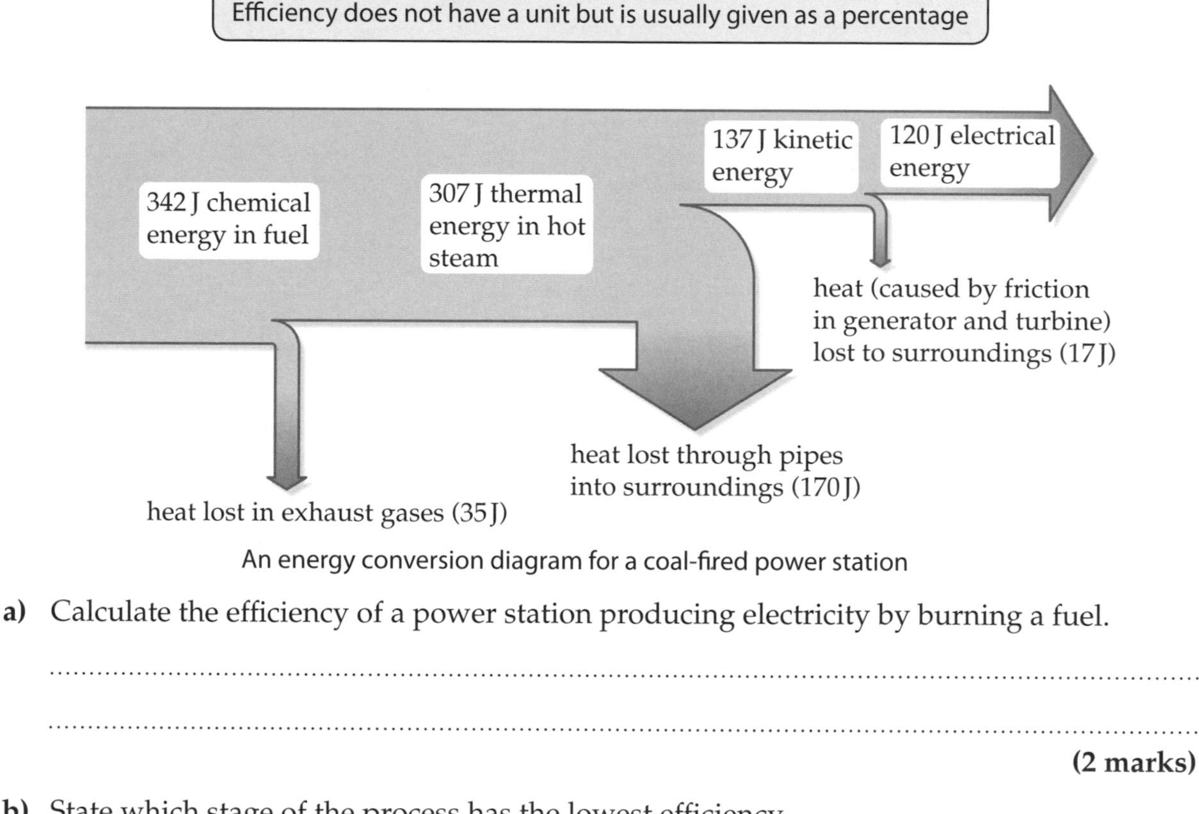

An energy conversion diagram for a coal-fired power station

a) Calculate the efficiency of a power station producing electricity by burning a fuel.

....

....

(2 marks)

b) State which stage of the process has the lowest efficiency.

....

(1 mark)

B-A*

3 The motor in a dishwasher has an efficiency of 20%. The motor produces 40 joules per second of kinetic energy. Calculate the amount of electrical energy that must be supplied to the motor each second.

Take care when you re-arrange the formula.

Electrical energy supplied .... J

(3 marks)

The Earth's temperature

1 The central heating system in a house has to supply 4 kW of heat to keep the house at a steady 20°C in the winter.

a) State the average power radiated by the house during the winter. Explain your answer.

The house radiates .... kW because at a constant temperature, power radiated = ...

(2 marks)

b) Suggest two different ways to raise the temperature in the house to 21°C.

The central heating could be ....

The power radiated by the house could be .... by ....

(2 marks)

D-B

2 A couple have a hot water tank in their house which is heated by an electrical heater. They find that their electricity bill is high because the heater is on all the time.

a) Explain why the heater has to be on all the time to keep the hot water at a constant temperature.

....

....

....

(3 marks)

b) Suggest **two** ways that the couple could reduce their electricity bill.

....

....

....

Think about the energy in and the energy out.

(2 marks)

B-A*

3 'Snowball Earth' occurred thousands of millions of years ago when the Earth was covered almost completely in ice and snow and the percentage of carbon dioxide in the atmosphere was much lower than it is today.

Explain why the average temperature of the Earth was much lower at that time than it is today.

....

....

....

....

(4 marks)

There are hints in the question, but use your scientific understanding to use them in your answer.

Physics extended writing 4

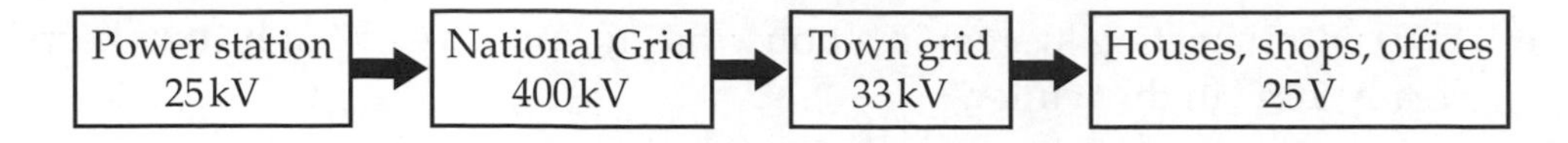

The diagram shows the changes made to the voltage of the electricity supplied by a power station to consumers some distance away.

Using the information in the diagram, describe how the voltage changes are made and the reasons for them.

(6 marks)

You will be more successful in extended writing questions if you plan your answer before you start writing.

The question asks you to describe the voltage changes made between power station and shops, offices and homes and why they are necessary. Think about:

- how the voltage is changed
- the scientific formula behind making the change
- why the voltage is changed in the National Grid
- why the voltage is changed before it reaches homes.

Make sure that your answer contains scientific principles – for example, you can include examples of the use of formulas.

Physics extended writing 5

These statements are about the predicted changes in global temperatures:

'Temperatures across the Earth are very sensitive to the amount of radiation absorbed and reflected by the ground.'

'The worry is that if we continue to burn fossil fuels we could cause a runaway greenhouse effect that results in the melting of the ice caps, which will only make the problem even worse.'

Discuss the points made in these two statements.

(6 marks)

You will be more successful in extended writing questions if you plan your answer before you start writing.

The question asks you to discuss the two statements on global temperatures. Think about:

- the balance between the rate of energy absorbed and rate of energy radiated in any system
- how this balance can be applied to the Earth – where does the energy come from and go to?
- what effect fossil fuels have on this balance.

Read *both* of the statements carefully and explain the points that they raise.

Make sure that your answer contains scientific principles.

Biology practice exam paper (allow one hour)

Edexcel publishes official Sample Assessment Material on its website. This practice exam paper has been written to help you practise what you have learned and may not be representative of a real exam paper.

Classification

1 The diagram shows a key. This key is used to classify organisms into kingdoms.

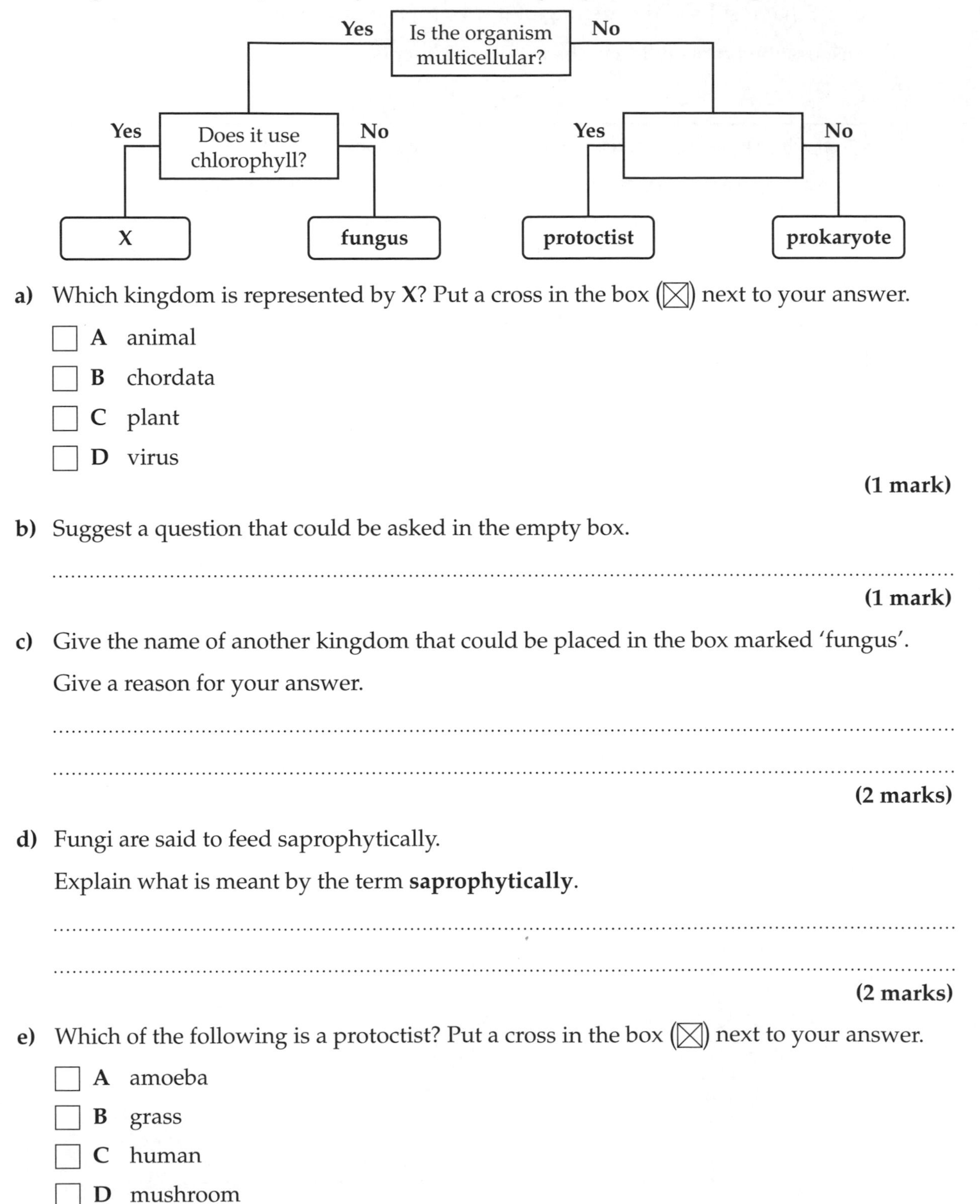

a) Which kingdom is represented by **X**? Put a cross in the box (☒) next to your answer.

☐ **A** animal

☐ **B** chordata

☐ **C** plant

☐ **D** virus

(1 mark)

b) Suggest a question that could be asked in the empty box.

....

(1 mark)

c) Give the name of another kingdom that could be placed in the box marked 'fungus'.

Give a reason for your answer.

....

....

(2 marks)

d) Fungi are said to feed saprophytically.

Explain what is meant by the term **saprophytically**.

....

....

(2 marks)

e) Which of the following is a protoctist? Put a cross in the box (☒) next to your answer.

☐ **A** amoeba

☐ **B** grass

☐ **C** human

☐ **D** mushroom

(1 mark)

f) Give one reason why scientists often class viruses as non-living.

....

(1 mark)

(Total for Question 1 = 8 marks)

Energy in the environment

2 A food chain shows the feeding relationships between organisms in a habitat.

grass → **slug** → **bird** → **fox**

a) Which organism is the secondary consumer? Put a cross in the box (☒) next to your answer.

☐ **A** grass

☐ **B** slug

☐ **C** bird

☐ **D** fox

(1 mark)

b) i) What is the initial source of energy used by the grass?

....

(1 mark)

ii) Only about 10% of the energy in one organism is passed to the next organism in the food chain.

State one reason why not all the energy is passed on.

....

(1 mark)

c) Sketch a pyramid of biomass for this food chain.

(2 marks)

d) Hedgehogs were introduced into this habitat. Hedgehogs eat slugs. Their skin is covered with spines, so they are not often preyed on by foxes.

Explain what happened to the numbers of foxes when hedgehogs were introduced.

....

....

....

(3 marks)

(Total for Question 2 = 8 marks)

Reactions times

3 Eight students in class 5W are doing an experiment to investigate their reaction times.

The first student puts on some headphones. A stopwatch is started and a sound is played.

When the student hears the sound, he presses a button to stop the stopwatch.

a) Which of these shows the path taken by the nerve impulse in this experiment? Put a cross in the box (☒) next to your answer.

☐ **A** ear – brain – muscle

☐ **B** ear – motor neurone – muscle

☐ **C** ear – sensory neurone – brain – motor neurone – muscle

☐ **D** ear – motor neurone – brain – sensory neurone – muscle

(1 mark)

b) The graph shows the data collected for the eight members of 5W in this experiment.

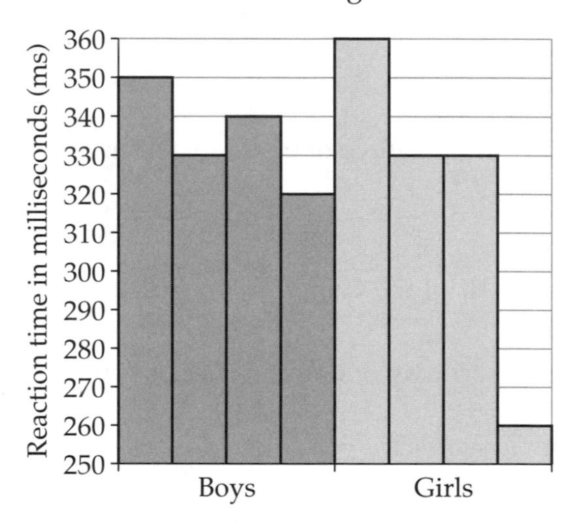

i) Calculate the average reaction time for the boys in the class.

Average reaction time .... ms **(2 marks)**

ii) Class 5W's teacher says that the average reaction time for the girls is 340 ms. Explain why the teacher uses this value for the reaction time of the girls.

....

....

(2 marks)

iii) The class make the following conclusion: 'Boys have a faster reaction time than girls.' Evaluate this conclusion.

....

....

(2 marks)

c) Explain why the results of this investigation could be different if the class had all drunk some coffee before doing the experiment.

....

....

....

(3 marks)

(Total for Question 3 = 10 marks)

Blood sugar

4 The pancreas and the liver are two organs involved in regulating blood glucose levels.

a) i) Suggest a food that a diabetic could eat to increase their blood glucose level.

....

(1 mark)

ii) Name the hormone that is released from the pancreas when blood glucose levels rise. Put a cross in the box (☒) next to your answer.

☐ **A** ADH

☐ **B** adrenaline

☐ **C** insulin

☐ **D** testosterone **(1 mark)**

b) People with diabetes have difficulty in regulating the level of glucose in their blood.

The risk of developing diabetes can be linked to a person's weight. The table shows the BMI values for people with different weights.

BMI can be calculated using the equation:

$$\text{BMI} = \text{weight in kilograms} \div (\text{height in metres})^2$$

BMI value	Weight
above 30	obese
25–30	overweight
18.5–25	normal
under 18.5	underweight

A man is 176 cm tall and weighs 86 kg.

Calculate his BMI and evaluate the risk of him developing diabetes.

BMI ....

....

....

(3 marks)

c) If there is excess glucose in the blood, it can be converted into a different compound. This compound is stored in the liver.

i) Name the compound that glucose is converted into.

....

(1 mark)

ii) Blood glucose levels fall in the blood as we exercise. Explain how the stored compound is used to return the blood glucose levels back to normal.

....

....

....

(2 marks)

d) The liver can be affected by a condition known as cirrhosis. Describe what is meant by cirrhosis, and give a reason why it occurs.

....

....

(2 marks)

(Total for Question 4 = 10 marks)

Chromosomes and genes

5 The chromosomes in our cells are made up of a series of genes. These genes determine our characteristics.

a) Where in the cell are chromosomes found?

....

(1 mark)

b) Complete the following sentence by putting a cross in the box (☒) next to your answer.

Genes exist in different forms that are called

☐ **A** alleles

☐ **B** chromosomes

☐ **C** DNA

☐ **D** gametes

(1 mark)

c) The condition cystic fibrosis is a recessive genetic disorder. The forms of the gene for cystic fibrosis can be shown as **F** (normal) and **f** (cystic fibrosis).

i) What is meant by the term **recessive**?

....

....

(2 marks)

ii) What is the genotype and phenotype for a person who is heterozygous for the cystic fibrosis condition?

genotype ....

phenotype ....

(2 marks)

d) Gaucher's disease is a recessive genetic disorder, which develops early in childhood.

A couple are going to have a baby. The woman has been tested and knows that her genes are **GG** – she is not a carrier of Gaucher's disease. The man knows that there is a history of the disease in his family, so thinks he might be a carrier.

By considering the man's possible genotypes, explain the phenotypes that could occur in their children. You may illustrate your answer with genetic diagrams.

....

....

....

....

....

....

....

....

....

....

....

....

(6 marks)

(Total for Question 5 = 12 marks)

Plant growth hormones

6 Plants produce a group of hormones called plant growth substances. These compounds act in a similar way to human hormones and control growth in the plant.

a) Gibberellins are plant growth substances.

Use the words in the box to complete the sentence about gibberellins.

cellulose	roots	starch	water

When gibberellins are produced, they help convert .... stored in the seed into glucose to help the plant to grow.

(1 mark)

b) Strawberries ripen naturally on a strawberry plant.

Describe **one** reason why farmers might use a hormone to ripen strawberries artificially.

....

....

(2 marks)

c) Auxins are plant growth substances.

Explain the effect that auxins have on the roots of plants.

....

....

....

(3 marks)

d) A student wants to carry out an experiment to see if plants grow towards the light.

He has two plants. He waters one and places it in the dark. He puts the other plant in a greenhouse, where it is watered every day.

The diagram shows the appearance of the plants after two weeks.

well-lit greenhouse

dark

He concludes 'Plants grow towards the light, because the plant in the greenhouse grew more than the one in the dark.'

His teacher says that the conclusion is difficult to justify as there were flaws in the experiment.

Explain why his conclusion cannot be justified from the experiment carried out, and how he could alter the experiment to allow a better conclusion to be drawn.

....

....

....

....

....

....

....

....

....

....

....

....

(6 marks)

(Total for Question 6 = 12 marks)

Chemistry practice exam paper (allow one hour)

Edexcel publishes official Sample Assessment Material on its website. This practice exam paper has been written to help you practise what you have learned and may not be representative of a real exam paper.

Hydrocarbon fuels

1 Oil is a fossil fuel.

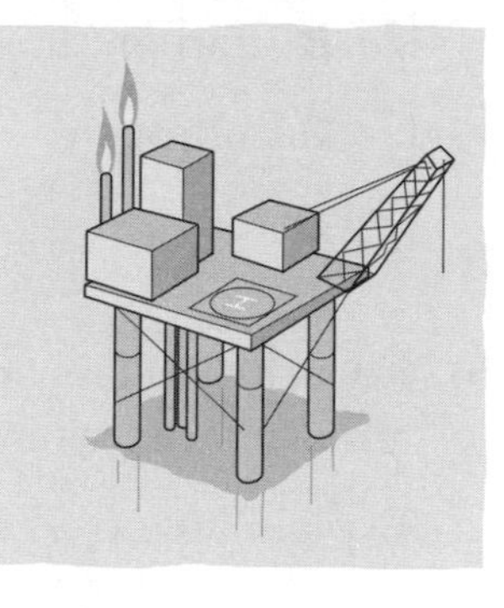

a) Which of these is also a fossil fuel?
Put a cross in the box (☒) next to your answer.

☐ **A** coal

☐ **B** paper

☐ **C** wax

☐ **D** wood

(1 mark)

b) The oil that we get from the ground is a mixture of different compounds.

These compounds are all hydrocarbons.

i) How are the compounds in crude oil separated from each other?
Put a cross in the box (☒) next to your answer.

☐ **A** chromatography

☐ **B** evaporation

☐ **C** filtration

☐ **D** fractional distillation

(1 mark)

ii) Explain what is meant by the word **hydrocarbon**.

....

....

(2 marks)

c) Propane and octane are just two of the hydrocarbons present in oil.

Propane can be used as a fuel. It burns to produce energy and new chemical products.

i) Complete the word equation showing the **incomplete** combustion of propane.

propane + oxygen → .... + water

(1 mark)

ii) Most of the time, propane burns through **complete** combustion.
Describe why complete combustion causes problems in the environment.

....

....

(2 marks)

iii) Balance the equation showing the **complete** combustion of octane.

C_3H_8 + ... $O_2 \rightarrow 3CO_2$ + ... H_2O

(1 mark)

(Total for Question 1 = 8 marks)

Gases in the air

2 The air that we breathe is a mixture of different gases.

a) Which of these gases is present in the **smallest** amount in air?
Put a cross in the box (☒) next to your answer.

☐ **A** argon

☐ **B** carbon dioxide

☐ **C** nitrogen

☐ **D** oxygen

(1 mark)

b) A teacher's class are doing a project on oxygen in the air. She tells them that the early atmosphere was mostly nitrogen and carbon dioxide.

Explain how oxygen was added to the early atmosphere.

....

....

(2 marks)

c) The teacher gives the class a sample of oxygen.

Describe a test that the students could do to show that the gas was oxygen.

....

....

(2 marks)

d) The teacher sets up an experiment to show the class how much oxygen is in the air.

She passes 100 cm^3 of air back and forward over some heated copper in a tube.

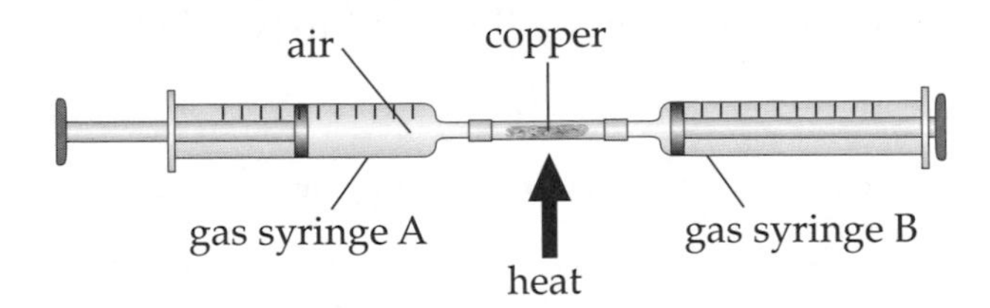

After passing the air back and forward over the hot copper four times, the volume of gas remaining was found to be 85 cm^3.

i) Using these results, calculate the percentage of oxygen in the air.

....

Percentage oxygen in the air .... %

(1 mark)

ii) Discuss reasons why this value is not the value that the class expected to get.

....

....

(2 marks)

(Total for Question 2 = 8 marks)

Zinc: a typical metal

3 Zinc is a metal. Zinc is obtained from zinc blende, which is found in rocks. Zinc blende is mostly zinc sulphide, ZnS.

a) i) What name is given to substances that occur in rocks, like zinc blende, from which the metal is extracted? Put a cross in the box (☒) next to your answer.

☐ **A** igneous

☐ **B** limestone

☐ **C** magma

☐ **D** ore **(1 mark)**

ii) Some metals, like gold, are found in the ground as elements, not as compounds. Explain why zinc is not found in the ground as an element.

....

....

(2 marks)

b) To extract zinc from zinc blende, the first step is to heat the zinc blende in air.

zinc sulfide + oxygen → zinc oxide + sulfur dioxide

In one day, 1.94 tonnes of zinc sulfide react with 0.96 tonnes of oxygen.

This produces zinc oxide and 1.28 tonnes of sulfur dioxide.

What mass of zinc oxide is made?

Mass .... tonnes

(1 mark)

c) In the second stage of the process, zinc oxide is turned into zinc.

i) What name is given to the process of turning zinc oxide into zinc? Put a cross in the box (☒) next to your answer.

☐ **A** combustion

☐ **B** neutralisation

☐ **C** oxidation

☐ **D** reduction **(1 mark)**

ii) There are two ways of completing this process. The first is to use electrolysis. The second is to heat zinc oxide with carbon. Explain **one** factor that might influence the method that is chosen.

....

....

(2 marks)

d) We currently use about 12 tonnes of zinc each year.

Scientists estimated that about 346 tonnes of zinc still remain in the rocks in the Earth.

i) Calculate how many years it will be before we run out of zinc.

Give your answer to the nearest year.

Time .... years

(1 mark)

ii) Describe **two** ways in which we can try to make our supplies of zinc last longer.

....

....

(2 marks)

(Total for Question 3 = 10 marks)

Hydrochloric acid

4 Two students are doing an experiment.

They are going to pass electricity from a battery through hydrochloric acid.

a) One student tries to explain to the other what the symbol on the bottle of hydrochloric acid means.

i) What does this symbol mean?
Put a cross in the box (☒) next to your answer.

☐ **A** corrosive

☐ **B** flammable

☐ **C** oxidising

☐ **D** toxic

(1 mark)

ii) Although hydrochloric acid is harmful, humans make this acid in our stomachs.

Give **two** reasons why this acid is important for humans to have in the stomach.

....

....

(2 marks)

b) When they do the experiment, one student knocks over the bottle of hydrochloric acid.

Some acid spills onto the bench. Their teacher pours some solid sodium carbonate onto the spill.

i) Describe the type of reaction that happens and an observation that could be made when the sodium carbonate and the hydrochloric acid mix.

Type of reaction

....

Observation

....

(2 marks)

ii) Give the name of the salt that is made in the reaction.

....

(1 mark)

c) The two students set up their electrolysis reaction, as shown in the diagram.

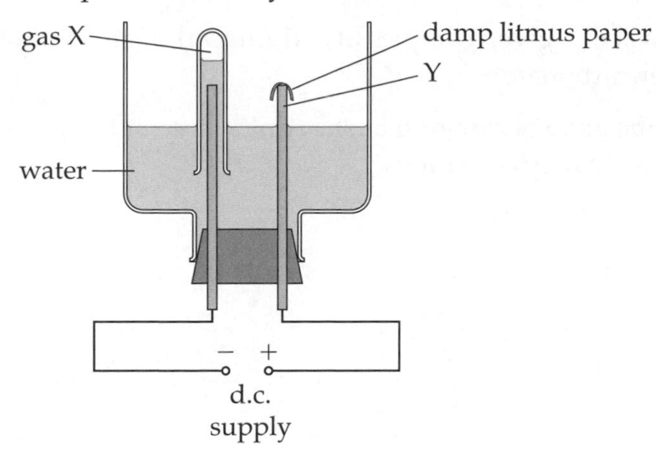

i) One student thinks that gas X is hydrogen. Describe a test he could do to show that the gas is hydrogen.

....

(1 mark)

ii) Describe and explain the observations the students would make at electrode Y.

....

....

....

(3 marks)

(Total for Question 4 = 10 marks)

Calcium carbonate

5 Calcium carbonate is found as both a sedimentary and a metamorphic rock.

a) i) Give the name of each of these calcium carbonate rocks.

metamorphic ....

sedimentary ....

(2 marks)

ii) Describe how sedimentary rocks are made.

....

....

(2 marks)

iii) Suggest why calcium carbonate is never found as an igneous rock.

....

....

(2 marks)

b) A student is investigating how metal carbonates decompose when they are heated.

She has been given samples of sodium carbonate, calcium carbonate, zinc carbonate and copper carbonate.

Describe an experiment that she could carry out to place these carbonates in the order of how easily they decompose.

....

....

....

....

....

....

....

....

....

....

(6 marks)

(Total for Question 5 = 12 marks)

Ethane and ethene

6 Two families of hydrocarbons are the alkanes and the alkenes.

Alkanes are saturated hydrocarbons, and alkenes are unsaturated hydrocarbons.

a) i) State the formula of ethane.

....

(1 mark)

ii) Draw the structure of a molecule of ethene.

(2 marks)

iii) Describe a test (with the result) that would allow you to tell the difference between ethane and ethene.

....

....

....

(3 marks)

b) Under the right conditions, ethene can react with itself to form a polymer called poly(ethene).

Explain how the properties of this polymer make it useful, but also lead to some problems with its use.

....

....

....

....

....

....

....

....

....

....

(6 marks)

(Total for Question 6 = 12 marks)

Physics practice exam paper (allow one hour)

Edexcel publishes official Sample Assessment Material on its website. This practice exam paper has been written to help you practise what you have learned and may not be representative of a real exam paper.

All the formulae you need are given on page 120.

Waves

1 A student fills a bath but leaves the tap dripping. The drips make waves in the bath water. A rubber duck bobs up and down.

a)

←0.12 m→

i) Use information on the diagram to help you to calculate the distance of the duck from the tap.

....

....

(2 marks)

ii) Which of these best describes why the rubber duck moves?
Put a cross in the box (☒) next to your answer.

☐ **A** Water from the tap pushes on the duck.

☐ **B** The waves transfer energy to the duck.

☐ **C** The filling bath gives the duck potential energy.

☐ **D** The duck floats on water.

(1 mark)

iii) The tap drips with a frequency of 2 Hz.
Calculate the speed of the waves.

Speed .... m/s

(2 marks)

b) He puts a plastic box upside down in the bath. The top surface of the box is just below the water.

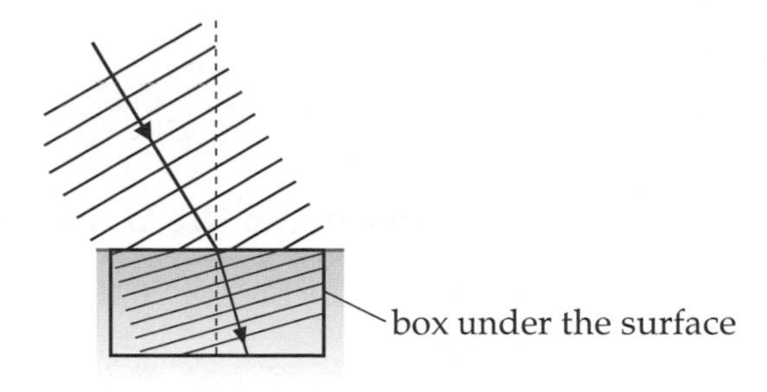

i) State the name of the effect shown in the diagram.

....

(1 mark)

ii) State what has happened to the speed and direction of the waves passing over the box.

....

....

(2 marks)

(Total for Question 1 = 8 marks)

Electromagnetic waves

2 a) Some people use an ultraviolet lamp to give themselves a suntan. When the lamps are on they usually also give out blue light. It is the ultraviolet light that tans the skin.

i) Which of the following statements is **not** correct?
Put a cross in the box (☒) next to your answer.

☐ **A** Blue light and ultraviolet rays are both electromagnetic waves.

☐ **B** Ultraviolet light carries more energy than blue light.

☐ **C** In a vacuum ultraviolet light travels faster than blue light.

☐ **D** Ultraviolet light causes tanning, blue light does not.

(1 mark)

ii) Explain why it is useful that the lamp gives out blue light as well as ultraviolet light.

....

....

(2 marks)

iii) Explain why it is important to wear dark goggles when using the ultraviolet lamp to tan the face.

....

....

(2 marks)

b) Some people say that the sale of ultraviolet lamps should be banned and that they should only be used in licensed tanning salons.

Discuss two reasons for agreeing with this statement.

....

....

(2 marks)

c) State one other use of ultraviolet light.

....

(1 mark)

(Total for Question 2 = 8 marks)

Waves and the Universe

3 Until 1965 the Big Bang Theory and the Steady State Theory were alternative explanations for observations of the Universe.

a) Which of the following is correct for the two theories?
Put a cross in the box (☒) next to your answer.

☐ **A** Both theories say that the Universe is expanding.

☐ **B** Only the Big Bang Theory says that the Universe is expanding.

☐ **C** Only the Steady State Theory says that the Universe is expanding.

☐ **D** Both theories say that the Universe began with a burst of energy.

(1 mark)

b) Cosmic Microwave Background (CMB) radiation was discovered in 1965 using a ground-based receiver. In the 1990s the COBE space probe investigated the CMB radiation. In 2001 the WMAP space probe was launched; its nine year mission was to map the CMB radiation across space.

The CMB radiation comes from every part of space.

i) State a reason why we cannot see this radiation with our eyes.

....

(1 mark)

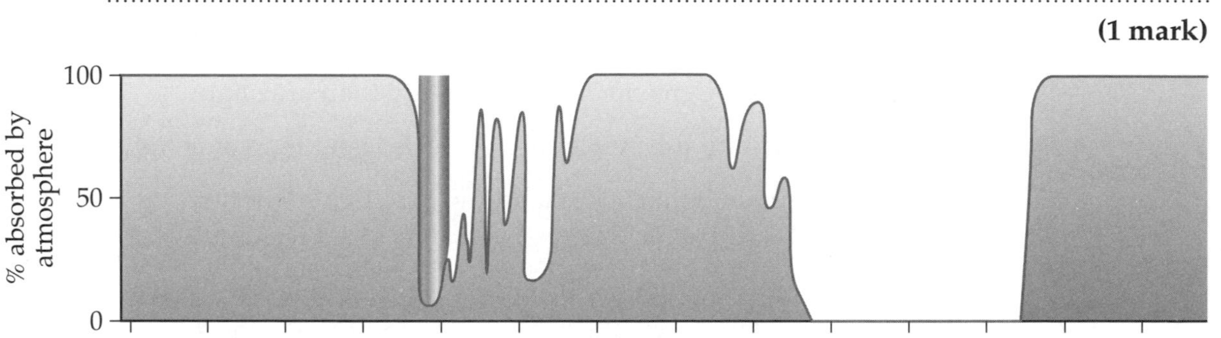

The absorption of electromagnetic waves by the atmosphere

ii) The wavelength of the CMB radiation is in the range 0.1 to 1 cm.

Explain why space probes are necessary to produce detailed information on the CMB radiation.

....

....

(2 marks)

iii) Measurements of the CMB radiation show that the average temperature of the Universe is about 3 degrees above absolute zero.

Suggest why the WMAP space probe has been placed in an orbit four times further from the Earth than the Moon and is shielded from the Earth and the Sun.

....

(1 mark)

c) Since 1965, the Big Bang Theory has become accepted by most scientists as being the best description of the formation of the Universe.

Explain why this has happened.

....

....

(2 marks)

d) All the stars we see in the sky today are younger than the source of the CMB radiation. The stages a star passes though during its lifetime depend on the mass of the star.

Describe **three** differences in the life cycle of stars similar to the Sun and stars that have a much higher mass.

....

....

....

(3 marks)

(Total for Question 3 = 10 marks)

Waves and the Earth

4 In 2011 a powerful earthquake occurred very close to the city of Christchurch in New Zealand. Scientists used data from seismometers to work out that the earthquake took place at a depth of 4.8 km. The first seismic waves (P waves) reached the surface directly above the earthquake 0.8 s after it happened.

a) State what type of vibration describes P waves.

....

(1 mark)

b) S waves arrive at the seismometer a few seconds after the P waves.

State the reason for the later arrival of the S waves.

....

(1 mark)

c) Calculate the speed of the P waves.

Speed .... km/s

(2 marks)

d) The actual speed of the P waves at the surface was probably less than your answer.

Explain why your answer is different from the speed at the surface.

....

....

(2 marks)

e) A weaker signal was received by the same recording station about ten minutes later.

Explain how this signal was received.

....

....

(2 marks)

f) Christchurch is close to the boundary between two tectonic plates.

Explain why this made an earthquake likely.

....

....

(2 marks)

(Total for Question 4 = 10 marks)

Generation and transmission of electricity

5 Small wind turbines are used to provide electricity for charging batteries for boats and caravans and in remote places. The spinning turbine turns the coil of a generator producing a current.

a) One type of wind turbine produces a current of 8 A to charge batteries at 12 V.

i) Calculate the power produced by the turbine.

Power produced .... watts

(2 marks)

ii) The same type of turbine can be used to supply electricity to the National Grid at 230 V by using a transformer.

Explain why a transformer can be used to change the output voltage of the turbine generator but not the output from a battery.

....

....

(2 marks)

iii) The transformer has a primary coil with 100 turns.

Calculate the number of turns on the secondary coil.

....

....

....

(3 marks)

b) Wind turbines are one source of renewable energy available on both a small and a large scale.

Discuss the advantages and disadvantages of wind, solar energy and hydroelectric power schemes for supplying electricity.

....

....

....

....

....

....

(6 marks)

(Total for Question 5 = 13 marks)

Energy and efficiency

6 A farmer uses a diesel engine connected to a generator to provide electricity to pump water up from a river to irrigate his fields.

a) The diesel engine is 40% efficient. In one second the diesel engine uses fuel with an energy content of 15 kJ.

i) Calculate the useful energy produced by the engine every second.

....

....

(2 marks)

ii) The engine wastes energy in the form of thermal energy. The engine is fitted with a radiator to transfer this thermal energy to the air around the engine. The radiator maintains the engine at a constant temperature.

Describe what would happen to the temperature of the engine if the radiator stopped working.

....

....

(2 marks)

iii) State a reason why the radiator of the engine is painted black.

....

(1 mark)

b) Discuss the energy transfers involved in using the diesel engine and generator to move water up from the river to the fields.

....

....

....

....

....

....

....

....

....

....

....

....

(6 marks)

(Total for Question 6 = 11 marks)

Formulae

You may find the following formulae useful.

wave speed = frequency × wavelength $v = f \times \lambda$

wave speed $= \frac{\text{distance}}{\text{time}}$ $v = \frac{x}{t}$

electrical power = current × potential difference $P = I \times V$

cost of electricity = power × time × cost of 1 kilowatt-hour

power $= \frac{\text{energy used}}{\text{time taken}}$ $P = \frac{E}{t}$

efficiency $= \frac{\text{(useful energy transferred by the device)}}{\text{(total energy supplied to the device)}} \times 100\%$

$\frac{\text{primary voltage}}{\text{secondary voltage}} = \frac{\text{number of turns on primary coil}}{\text{number of turns on secondary coil}}$ $\frac{Vp}{Vs} = \frac{Np}{Ns}$

Periodic table

Key

relative atomic mass
atomic symbol
name
atomic (proton) number

1	2											3	4	5	6	7	0
							1 **H** hydrogen 1										4 **He** helium 2
7 **Li** lithium 3	9 **Be** beryllium 4											11 **B** boron 5	12 **C** carbon 6	14 **N** nitrogen 7	16 **O** oxygen 8	19 **F** fluorine 9	20 **Ne** neon 10
23 **Na** sodium 11	24 **Mg** magnesium 12											27 **Al** aluminium 13	28 **Si** silicon 14	29 **P** phosphorus 15	31 **S** sulfur 16	35.5 **Cl** chlorine 17	40 **Ar** argon 18
39 **K** potassium 19	40 **Ca** calcium 20	45 **Sc** scandium 21	48 **Ti** titanium 22	51 **V** vanadium 23	52 **Cr** chromium 24	55 **Mn** manganese 25	56 **Fe** iron 26	59 **Co** cobalt 27	59 **Ni** nickel 28	63.5 **Cu** copper 29	65 **Zn** zinc 30	70 **Ga** gallium 31	73 **Ge** germanium 32	75 **As** arsenic 33	79 **Se** selenium 34	80 **Br** bromine 35	84 **Kr** krypton 36
85 **Rb** rubidium 37	88 **Sr** strontium 38	89 **Y** yttrium 39	91 **Zr** zirconium 40	93 **Nb** niobium 41	96 **Mo** molybdenum 42	[98] **Tc** technetium 43	101 **Ru** ruthenium 44	103 **Rh** rhodium 45	106 **Pd** palladium 46	108 **Ag** silver 47	112 **Cd** cadmium 48	115 **In** indium 49	119 **Sn** tin 50	122 **Sb** antimony 51	128 **Te** tellurium 52	127 **I** iodine 53	131 **Xe** xenon 54
133 **Cs** caesium 55	137 **Ba** barium 56	139 **La*** lanthanum 57	178 **Hf** hafnium 72	181 **Ta** tantaium 73	184 **W** tungsten 74	186 **Re** rhenium 75	190 **Os** osmium 76	192 **Ir** iridium 77	195 **Pt** platinum 78	197 **Au** gold 79	201 **Hg** mercury 80	204 **Tl** thallium 81	207 **Pb** lead 82	209 **Bi** bismuth 83	[209] **Po** polonium 84	[210] **At** astatine 85	[222] **Rn** radon 86
[223] **Mn** francium 87	[226] **Ra** radium 88	[227] **Ac*** actinium 89	[261] **Rf** rutherfordium 104	[262] **Db** dubnium 105	[266] **Sg** seaborgium 106	[264] **Bh** bohrium 107	[277] **Hs** hassium 108	[268] **Mt** meitnerium 109	[271] **Ds** darmstadtium 110	[272] **Rg** roentgenium 111	Elements with atomic numbers 112–116 have been reported but not fully authenticated						

** The lanthanoids (atomic numbers 58–71) and the actinoids (atomic numbers 90–103) have been omitted.*

The relevant atomic masses of copper and chlorine have not been rounded to the nearest whole number.

Final comments

Here are some other things to remember in your exam.

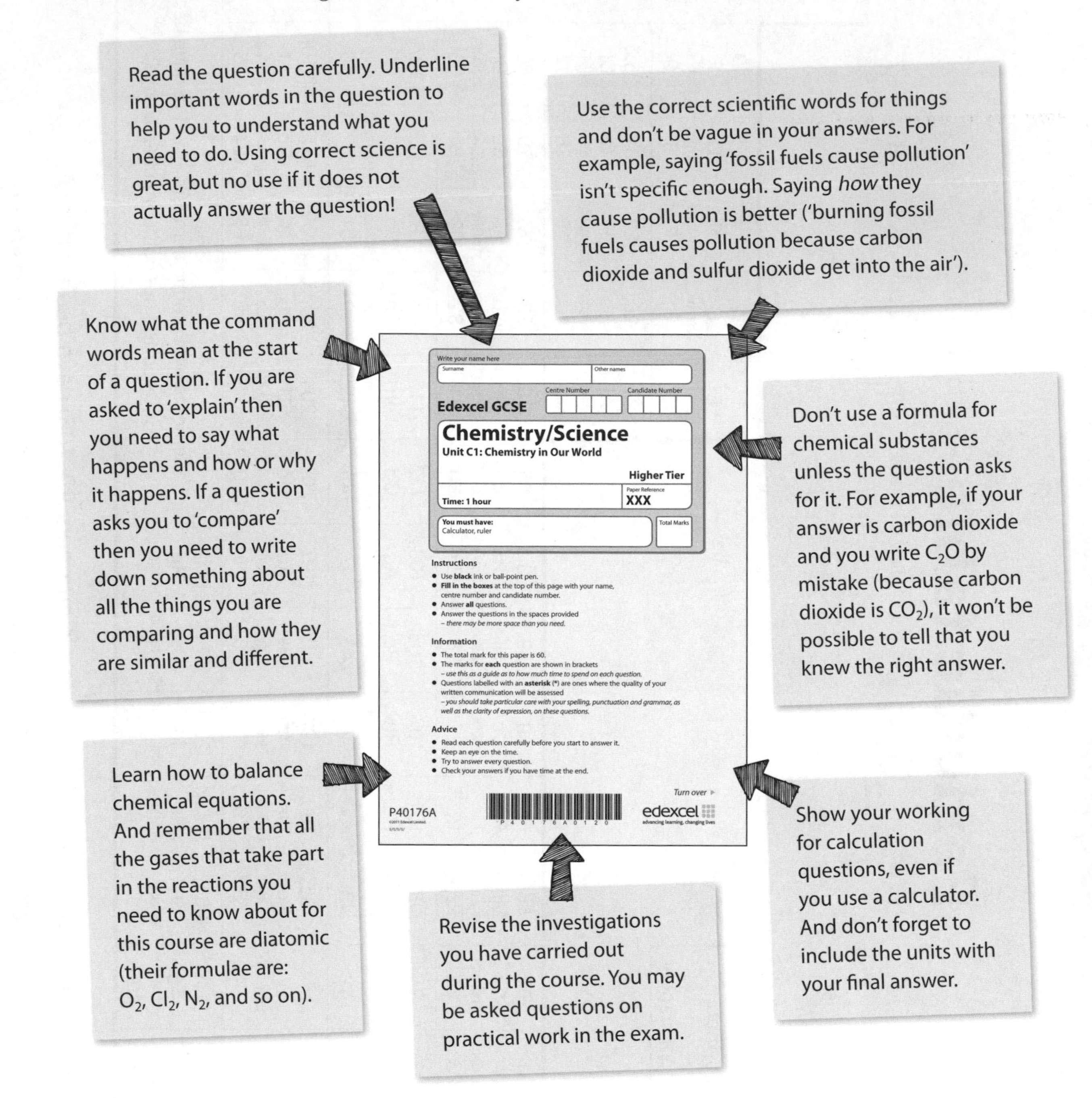

Using formula triangles

There will be a formula sheet in the exam, so you do not need to memorise equations, but you do need to be able to rearrange them.

If you cannot remember how to do this, you need to memorise the formula triangles given with formulae in this book. For example, $P = I \times V$ will be given in the exam paper. If you need to work out the voltage (V), cover up the V on the formula triangle. This will tell you that you need to divide P by I to get your answer.

$P = I \times V$ (given in exam)

This can be rearranged as:

$V = \frac{P}{I}$ or $I = \frac{P}{V}$

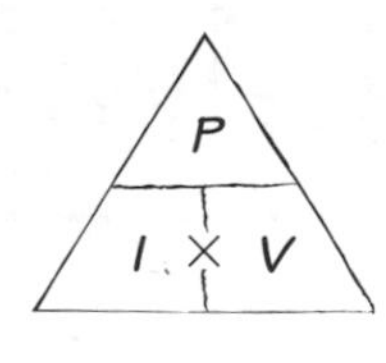